I0043821

J. G Goodchild

Book Tephi

J. G Goodchild

Book Tephi

ISBN/EAN: 9783743471870

Manufactured in Europe, USA, Canada, Australia, Japa

Cover: Foto ©berggeist007 / pixelio.de

Manufactured and distributed by brebook publishing software (www.brebook.com)

J. G Goodchild

Book Tephi

THE
BOOK OF TEPHI

THE BOOK OF TEPHI

THE BOOK OF TEPHI

BY

J. A. GOODCHILD

AUTHOR OF

"SOMNIA MEDICI," "THE TWO THRONES,"
"MY FRIENDS AT SANT' AMPELIO," ETC.

SID CO NEM

NEM CO DOMAN

DOMAN FO NIM

NERT HI CACH

"He is cursing in rhyme, and with two assonances in every line of his curse."
The Crucifixion of the Gleeman, BY W. B. YATES

LONDON

KEGAN PAUL, TRENCH, TRÜBNER & CO. LTD.

1897

PREFACE

In 1886 I published a fragment of this tale which some of my reviewers then invited me to complete. I have been unable to accept this invitation earlier owing to my own ignorance of the story as told by the Irish bards; and these, so far, give me little help between the departure of the "sea-king's daughter from over the sea" from Taphanes and her arrival in Ireland; though I fancy that eventually something might be gleaned upon this head from other Celtic sources, particularly those in which the name Inogen or its congeners appears. My own rough and erroneous reproduction of the main features of a story which has deeply influenced the national, clerical and literary history not merely of Celtdom, but of all non-Sclavonic Europe, is chiefly based upon the excellent modern translations of Messrs Standish O'Grady, Whitby Stokes, and others; whilst I must recognise the claim made by Gillariach, the crouchbacked, O'Clery, to kindly remembrance for preserving certain important details which would otherwise probably have been lost.

Mark well the imagery in the following imaginary passage from a discourse of a tattered and shorn disciple of Mog Ruach to a scanty but appreciative audience. It is taken from that sermon which he preached under the stars of a frosty Samhaim, being in soreness of body, and in very great bitterness of soul under the cursings of St Maelruan, and of the holy bishop Magnenn.

" *Ye that would still hear the wisdom of Semias, servant of the Holy, which he learned of Rudrofheasa, know how the common amongst you say that there be many gems in the pool of Crotta Cliath, and indeed your saying is a true one. Also ye call that pool the Lake of the Dragon's Mouth, and wherefore ?—It was in that pool that Ternog's nurse saw the great salmon which St Fursa cursed for a dragon into its mud.— Now, I swear unto you that this same dragon shall carry St John upon his day when he rideth to avenge his brother John Baptist upon the female saints of Eriu. On that day's eve is Fian Cinged born under the Brat Baghach. Threescore and ten stars are counted to it. Yet, oh my son, beware the black fourhorned moon which hath wings as hands, for thou art tender. Nevertheless, if those brethren be near, thou art safe with thy thousands upon Roth Ramach when thou wieldest the threefold besom. I see the*

*slender pillar to whose bolts men are blind. He that
heareth is deafened. Him that they seek, is dead.
Thus must my White Star diminish the red moon
and the third of the birds of prey. Lo, herein is the
wise teaching of Morfessa of Fal, and of Uiscias which
he taught in Tasiac Tuathaib upon the field of Mell.
This is that lore which Cesair daughter of the Great
King gathered of Ernmais in Egypt when she fled
from the flood and rested ere the ships were burned at
Belgadan. Hereof she instructed Mac Indoge before
she entered the sacred treasurehouse. Well do ye know
these things, and because of them shall Magnenn and
Maelruan of Tamlacht be hurled into your lake, and
Dil, the darling of my heart, swim upon Masbuskala
to destroy them. Yea, let curses of mighty Ollams and
Anrads, and my own curse which is less worthy, rest for
ever upon all that call the blackmaned heifer " sow " or
" serpent " ; and may her rugged one with the tusks of
his fork root up their graveyards, that their dry bones
may be foul beneath the sun and lie upon the heap for
ever."*

Upon such bottom for dragon or salmon lie objects
strongly refractive to starlight, though dark under the
candelabra of Pontiff or Kaiser. Experts are no
doubt right in referring them to the Fata Morgana,
but have not tested them with X Rays at present.

The commons still value rough specimens above coral and stoneware penates of nature and art, but I trust that few modern depreciators of Celtic moonstones will accept the suggestion of Irenæus, and the author of the "Testament of the Patriarchs," and expanded by many subsequent writers, that they are the produce of the Swart Sow and Malemantus of Dan. I may remark here that the general argument of the latter writer is against Levi, patron-patriarch of Peter and Patrick, rather than of John and Pelagius.

I am far too ignorant to analyse them. My own specimens are here, much dulled by my fingering. If they be pebbles irridescent with scum, they may be cleaned and reported upon by the mineralogist. If St Fursa is of her original opinion, she should get St George to help her to look after them : but IF the Great Salmon of Ollamhaba was indeed seen by Ternog's nurse, by the aid of Ruacha Aodhècis, many of its ova are hatched already, and the remainder lack but twenty-five years of their fullest term.

J. A. GOODCHILD.

June 23rd, 1897.

THE BOOK OF TEPHI

CHAPTER I

(1) Tephi, born in the House of the High Ones,—
    (Princes of Zion,
    Zion loved of the Lord,—home of the House
      of our God,)
    Daughter of David, shepherd in Judah,—(Tribe
      of the Lion)
    Queen over Bethel and Dan,—where they be
      scattered abroad.

(2) Is not the Word made sure?—We are spread
      forth in alien places.
    Fire that was kindled in wrath—burns to the
      uttermost Hell.
    Cry in the night oh Judah,—Thy wise men
      covered their faces.

A   1

Howl for thy young lions slains,—princes led
captive to Bel.

I, even I am left,—to cry from the uttermost
region,—

(Far off isles of the West,—home of the
remnant of Dan,)

Sown as a thistle on earth is Jacob,—the names
of us legion.

Tongue of the Hebrew fails,—shall not be
spoken of man.

Isaac is ploughed in his furrows,—before the
Lord in this season

Water the tender plant,—twig of the loftiest
shoot.

How is the cedar left bare—in its boughs was
corruption and treason,

Crown of it bended to Baal,—serpents devour-
ing its root.

Rest for the flock of the Lord—was not found in
the shade of the cedar.

Broken it lies. It burns.—Yea, as a thorn
'neath a pot.

Kidlings are seething therein—shot down by the
archers of Kedar.

Foemen are warmed thereby,—fire of its
furnace is hot.

Children of Edom dance,—yea, leap in the place
which is Holy.

Bethlehem boweth in chains,—trodden as
clay in the mire.

How are our walls broken down,—that the pride
of our mighty is lowly.

Yea we wander 'mid stones,—deserts of thistle
and briar.

(3) I, that am old was young,—but my heart ran
down into water,

Hearing battle and strife,—terror that riseth
by night,

Princes and warriors stricken,—fallen like sheep
unto slaughter ;

Women's wails in the streets,—outside the
clamour of fight.

How are the nobles fallen !—Yea, they were
strong, they were ruddy,

Fat with the firstlings of flocks,—strong with
the strength of the vine.

Now are they white with famine,—their garments
   of purple are bloody ;
Meat, is flesh of the child.—Blood of our
   people is wine.
These were as water spilled—on the ground
   before Nebuchadnezzar
Drops that the dogs licked up,—Have they
   not gathered and fled.
Leaving the women and babes,—Chaldæans
   should slaughter at pleasure.
I that was babe of the Kings—trembled alone
   by my bed.

(3) Yet one came thither unchid, to the place of
   the women he passed,
Feared of the king and hated, his hour had
   come at the last.
In the room of the sire, the prophet, the prisoner
   none might heed
Came through the wasted harvest to gather the
   chosen seed.
Sternly he bade me to follow.   I dared not look
   in his face

As he led me by secret ways to a cave 'neath
the Holiest Place.

Here was my one sure hold, and I dreaded it
not for the dark,

But I knew the fear of the Lord, I knew that His
holy Ark

Was near and I trembled for these, and I ate
the water and bread

Of affliction full three days wherein I dwelt as
the dead,

Till I heard the voice of Baruch smite from
the opened roof

"The foe is gone from the gates, and the path
of our way made smooth."

Then forth in the veil of smoke from the ashes
wherein she weeps

We passed through the walls of Zion, her
palaces fallen in heaps.

Look, cry aloud for she slumbers,—dreaming a
dream that awakes not;

Weep, tear thy garments in shame,—ashes and
dust on thy head.

Yea, though the wilderness howl,—yet the voice
of Jerusalem speaks not ;
Mourn for her exiles, mourn,—none break the
rest of her dead !
Where is the House of the Lord?—Desolation
and mourning and sorrow !
Where is the place of the King?—Torrent-
gash sun-scorched and brown.
River of rocks, burnt bones !—There the lizard
shall see him the morrow,
Scorpions find them a place,—conies make
nests for their own.

CHAPTER II

(1) *Tephi addresseth her sons, and telleth of her going into
Egypt ;* (2) *she prophesieth blindness on Joseph and
Judah ;* (3) *she dwelleth as Pharaoh's daughter at
Tahpanes ;* (4) *Baruch heareth of the road to Tarshish ;*
(5) *the Prophet prophesies against Egypt.*

(1) My children remember Zion.   Moreover I bid
you to mark
That the word of the Lord is holy, though His
purpose therein be dark.

Ye know how we came unto Mizpah, and trusted
    in peace to dwell
With the servant of God that was slain there. It
    needs not of this to tell;
But of this my sons take heed, shall not your
    hearts understand
How the Prophet of Zion prayed that our steps
    might be stayed in the Land?
Shall ye not read in His book of the hope of our
    rest undone
Of Ismael's fraud, of the tumult and flight, and
    of Shuphan's son
And how we went into Egypt?

(2)   Nay, Joseph shall long be blind,
An ox that sleepeth at midnight, and Judah
    couched as a hind.
The lion hath fled from his lair. The ox hath
    wandered astray
Till the dawn of the East be red, and the night of
    the North be grey,
In the night shall no man know them, or the
    signs that be left to show

Where the shepherd keepeth the ox, whilst the
   lion is couched full low.
Not by the banks of Jordan, not on the Holy
   Hill
Are Ephraim's feet till his furrows be ploughed
   unto Yahveh's will.
Bethlehem's field is empty.   The shepherd
   follows astray.
Hear ye my words, oh my sons, for the Isles shall
   await the day.
Tephi, I was but weak, a little thing in men's eyes,
A tender twig of the cedar, yet sheltered of
   prophesies.
The Prophet of God revealed this.   Is not his
   speech made plain?
He came to root and destroy.   He went forth to
   plant again.
In our fields he found no vineyard, on our pastures a wasted soil,
No place for the shade of cedars, no depth of
   the earth for oil.
Till the Land be fed by the Goim,* and the tale
   of their slaughters told

* Nations.

The days shall be slowly numbered, and the hope
of the hills wax old.

(3) I was led as a slave into Egypt, as a captive to
Pharaoh's hand
For the will of the son of Kareah rested still on
our band,
But the heart of Pharaoh was softened.   He
gave us a resting place.
As daughters we stood before him, and the
Prophet of God found grace
To lead us unto Taphanes, henceforth amongst
men to be
Jehudia, House of the daughter of Judah, mind-
ful of me
Unto the ending of days.

(4)   Therein a space was our rest
Till Baruch the scribe found tidings out of the
Isles of the West
That the ways unto Tarshish were open, the ships
of Javan afar,

And vessels of Tyre went forth on the left of the
    raclen's* star
From the tongue of the sea to Melcarth's porch
    of the setting sun,
Whence Northward and West they sailed till the
    Island of Towers was won,
On its righthand Bregan and Eber, on its left
    that water whose bound
Is the Promise of God, wherein His purpose
    shall yet be found.

(5) Then the Prophet prophesied greatly of wrath
    and of woe to come
Upon Misraim's king and people, and all that
    made Cush their home.
Weak and poor shall it be. Three kings shall
    come from the East
Nimrod, Madai and Elam to break down the
    sacred beast.
Javan and Chittim thereafter from the islands
    shall issue forth
To rule the rivers of Egypt and bear their spoils
    to the North,

           * Merchants.

Tursi and Roumi shall reign over these with an
    iron yoke
Till the gateway of Heaven be opened, and the
    fetters of death be broke ;
Yet the land shall be filled with trouble, lamen-
    tation, weeping and pain,
Though the Prince of Peace be born, and be
    lifted on high to reign
On the holy Hills ; for Sheba and Dedan shall
    overflow,
And across the broad Euphrates the moon shall
    arise in woe ;
As blood shall it shine from the world's high
    roof to its western gate,
A crescent that never filleth, and the Star of Peace
    shall it hate
Till the night be wellnigh ended ; and ships
    come out of the West
Whose mouths are as stinging serpents, and fires
    are within their breast ;
Yet the angels of God are with them.   The Rolls
    of the Law they bear,
The spirit of peace is with them, and the promise
    of peace they share.

Then Egypt shall be as water, Yet now shall the
 Nations rise,
And the books be opened upon them, yea, even
 in all men's eyes,
Of the wrath and the promise of Jacob, his sons
 be purged of their guilt,
The ways of the King be open ; and that house
 of our God be built
That shall never henceforth be shaken.
     These things be gravèd and set
In the lime by the kilns of Pharaoh. Their
 place shall be hidden yet.
Therewith is my story written, and carved on
 stone by the scribes
Are secrets of things which shall be, and the
 names of eleven tribes
At the end of their days appointed, but Judah
 goes thither and fro
As a stricken lion in the pit till the hour of the
 final woe.

CHAPTER III

(1) The sisters of Tephi desiring to remain in Egypt die there; (2) A vessel of Tarshish cometh into one of the mouths of the Nile; (3) Tephi goeth from Taphanes, but is anointed before her going; (4) her prophecy thereupon. *

(1) My sisters ye mourned not for Zion, though
    short was your day and sad,
Ye loved the fleshpots of Egypt, and marvelled
    my soul was glad
That the time of our voyage drew nearer. Ye
    longed with her gods to stay,
And the Angel of Death drew sword and both
    were slain in a day.

* When writing this part of my tale, my ignorance of the details of the story told by Irish writers led me into an injustice to Maacha and Bathba the sisters of Tephi. The former is said by them to have fallen whilst encouraging her sister's troops in the wing commanded by Nuadh at Moytura, but there are many errors and omissions in this work which would require far more skill and patience than I possess to rectify, in my endeavour to repair the neglect into which the tale has fallen. All my readers will however have caught one glimpse at least of these three weeping queens in the barge of King Arthur, as they bear him away to await his time and their own.

Then the servants of idols bound ye in aloes
　　and spice and myrrh,
And we laid you amongst the heathen, but not
　　in their sepulchre.
Baruch hath written your names on the wood,
　　and o'er either face
Skilled workmen moulded the gold where ye wait
　　in your resting place.
I might not weep.　Ye had sinned.　Upon
　　Egypt's sin was your love ;
And the cry of the Man of God drew down His
　　wrath from above.

(2) Now a ship drew near into haven, a ship from
　　the far-off seas,
Whose pilot was child of the Dannites, whose
　　sails had filled to the breeze
In the boundless river of God.　Returned from
　　the storehouse of tin,
It had weathered the sea of storms, and the
　　waters that rage therein.
Her tin she sold to the founders of brazen
　　vessels, and lead

That was cast in bolts for the slingers ; with many
    tires for the head
Of the locks that I knew too well, of the tresses
    that shimmer fire
Which flickers before men's eyes and fills their
    hearts with desire ;
And amber from wizard lands at whose dread
    the Lochlann mocks
When he sails his hidebound boat through the
    sea of the floating rocks,
Whence monsters with horns arise to behold the
    sun lie red
On the lap of the sea by night, nor reigns he
    at noon o'erhead.
Swiftly they loaded the ship with the good things
    out of the land,
Rich garments, and potter's vessels, and arms for
    a chieftain's band,
And beads of glass for the women, and oil and
    almonds and spice,
And gold of the cunning workmen, and food
    with their merchandise ;
Till we 'scaped in the night from Pharaoh, but
    hid in the field that day

Whilst the hand of the Lord held back the
watchmen that barred our way.

(3) We were five that rode upon asses, and five by
the mules they led
Whereon were the things brought forth from the
House of the Lord when we fled,
The stone of Jacob our father, the Seat wherein
Yahveh dwells
Upon sacred things whereof the Book of the
Prophet tells ;
And the signs of my father David, on whom was
the promise stayed
Bright as the crown of the dawn, deep as the
midnight shade,
Strong as the purpose of God when he fashioned
the land from the sea,
A hope for the sons of Adam, that the chosen of
Him should be
A King over men for ever ; yea, unto the Lord's
own day
When the land shall be broken in dust, and the
sea shall vanish away.

Upon me was that promise fallen. For me was
    the Prophet's toil.

He had signed me with David's signet, anointed
    mine head with oil.

He had set mine hands to the harp; he had
    bidden me hold the spear;

The buckler was girt to my bosom, and Baruch
    and he drew near

To set my feet upon Bethel, the Stone that is
    seen this day

That my seed may rest upon it where'er it is
    borne away,

And its promise be sure beneath them, strong
    to uphold their throne.

Though the builders cast it aside, it shall never
    be left alone.

These things we did at Taphanes ere we fled to
    the haven of ships,

And the spirit of God came on me; His promise
    rose to my lips.

I spake, and I bade go forward, and the sons
    of the Lord obeyed,

And the Prophet of God bowed down, and
    this was the song that I made.

B

(4) As a seed in a desert amongst thorns—
    I am fallen. I am blown by the wind.
In thy garden, in thy pleasant field, beloved,—
    Is no water, is no rest that I may find.
Bel hath broken down thy cisterns and thy
       founts,—
    Esau cast his sum upon thee in thy woe.
Misraim's night is as a darkness to be felt,—
    Follow ye with me the sun where'er it go.
Follow after, follow after, my beloved,—
    Follow after by the pathways of the deep.
Leave the cloud of midnight thick upon this land.—
    Go before the sun that riseth out of sleep.
Plant me far upon the far green hills.—
    Ye have poured a living oil into mine heart,
The waters of the sea shall gird me round,—
    As the armour of the shield when I depart.
My children hearken to an holy harp,—
    As a certain sign of promise this shall be.
The spear within my right hand will I keep,—
    As the sceptre of the billows of the sea ;
And the lion of my signet is a sign,—*

---

* Tephi is alluded to by an early writer as the "blackhaired
heifer, the dark heaven-sealed chief, the lion."

Yea he roareth unto them that dwell afar.
And the name of God engraved therein shall
    cry,—
In the darkness as a light and guiding star.

CHAPTER IV

(1) *Tephi goeth from Egypt and cometh unto Carthage;* (2)
The Prophet maketh the Burden of the city; (3) *A
storm cometh out of the desert and the ship is driven
away until they come to a river in a strange country.*

(1) ON a moonless night and a cloudy we shipped
    and we passed away
In the veils of the Lord from Egypt. The breath
    of His mouth was our stay
Three weeks in our sails to westward. Thus
    favour was in the eyes
Of the men of the ship upon us, and I talked
    with our pilot wise,
Buchi the son of Helek, whose marvellous words
    were truth
He had gathered in many waters, an old man
    now from his youth,

Who in barks of Dan and Javan had raised up
    sails as a boy
For the sons of some that Ulick son of Liart
    brought back from Troy.
I heard of the painted talking birds in gardens
    with fruits of gold ;
And fish islands spouting fountains ; and one
    terrible tale he told
Of a giant that dwelt amongst trees, and descend-
    ing rended in twain
Three Miledhs * that sought him with target
    and spear, but in fight were slain.
In his hairy hands were they twisted, yea, as a
    stalk that is bent
On the myrtle ere it be gathered, so were they
    broken and rent.
Thus we came to the Kirjath Hadtho, and
    moored at the long fair wharf
Whence Ham and his camels athirst seek the
    treebuilt homes of the dwarf,
And beheld the Bozrah above it, yet set not
    our feet therein,

---

* Warriors, Milesians (Milites).

For Canaan, Phut and Lubim be wholly bound
  unto sin ;
And Buchi spake of their princes, and how when
  a Shophet died,
His wives were brought to his burning, his slaves
  to be crucified ;
Of Ashtaroth and of Tanith, queen harlots of
  cruel name
Whom the Fœni brought from the East ere into
  their land they came,
And of Baal whom Yahveh hateth. He dwelleth
  amongst you still.
Ye sons of Erin, I know ye. I know that your
  hands work ill.
Root up the groves from among you. Cast down
  his seats on the tors.
His fires are destroyers of gladness, his feasting
  my soul abhors.

(2) Hear ye, hear ye, that which he spake, the
  Prophet of God
When he stood betwixt Baruch and Buchi and
  stretched on that land his rod.

" Baal shall be broken," he said, " Yet he shall
    rise as the sun,
Red and gold is his rising.   Swiftly his course he
    shall run,
Unto the isles of the West, unto the uttermost sea,
Unto the land of the Sikels surely his border
    shall be.
Nemidh kneeleth his camel, fat is he waxen, and
    full.
The wealth of many waters hath swollen the hide
    of the bull.
A son is born him in season.   Yea, as a tiger's whelp,
To the West doth he leap, to the North, to the
    South.   There is none that may help.
By his teeth are men slain, in his claws they are
    rent, and the chief of his prey
Are the cubs of the wolf who mourns not, but
    ever croucheth at bay.
In the blood of her cubs he is sick, he is blind,
    he is drunken, he falls.
Hear it, ye gods of the heathen.   Hear it, ye far-
    stretching walls.
The wrath of the she-wolf is sated.   Your place
    is spread as a plain.

Your altars of blood are cast down.   Your fires
    unto Baal are vain.
The Tusci and Roum burn you.   Their host is
    come out of the North,
As on Nimrod and Assur and Edom and Tyre,
    the curse hath gone forth,
Thy sons shall be few and scattered, thy daughters
    carried to shame,
Thy walls be broken for ever, thy temples set to
    the flame."

(3) The West was blood as he spake.   The sky was
    black on the land,
The blast of a furnace sped from the trackless
    ocean of sand
Bearing the wrath of Baal, and smote on the
    Prophet's mouth,
But the hand of the Lord was with us to turn our
    way from the South.
Our sails were rent, and the men of the vessel
    cursed us by names
Of their gods, but feared the Prophet who called
    out of heaven its flames,

Fire and hailstones and thunders, and hills from
  the tossing sea ;
But I stood beside him and feared not, for helpers
  of heaven were we.
Seven days did I stand beside him with Buchi
  the pilot of Dan,
And the eyes of the Fœni hated, yet hoped in
  the waveworn man
And the child and the Prophet only ; for Baruch
  kept watch below
By the Stone and prayed upon it to comfort my
  women's woe.
Whither we went we knew not, yet Buchi stood
  by the helm,
Whilst the waves sped hungry after, but dared not
  to overwhelm
The Prophet of God, and the daughter of hope
  who stood by his side,
That the name of the Lord might stand, and his
  promise be magnified.
But the Fœin bowed down and blessed us when
  now on the seventh day
The sea was at Sabbath stillness, and we entered
  a little bay

By the mouth of an unknown river that ran from
 East unto West,
And lay tawny beside the shore where we anchored
 and lay at rest.

CHAPTER V

(1) Then the men consulted together, and marvelled
 upon that spot,
And Bœdan the son of Buchi was chosen of them
 by lot
To lead our skiff to the shore, and find of the
 folk thereby
What hap had fallen upon us, and whither our
 course should lie.
Now Bœdan brought us a man that they caught
 in a bushy field,
On his head a brazen helmet, on his left arm a
 broad round shield,

At his thigh a short stiff falchion. His feet were
    mired in the clay

Of the marsh where Bœdan traced him, and
    caught and brought him away.

Now the man bent not before us, but gazed with
    a steadfast eye

On our engines of war and weapons, and spake
    no word of reply

Unto Buchi who spake all tongues, till the gaze
    of the Prophet fell

Upon him compelling and silent, and then he
    spake full well

In a tongue that the Sicans use. " I come from
    the she-wolf's hold

Nigh at hand on the river, to seek a sheep of my
    fold.

I am very wroth, ye Fœni. I am wroth with the
    son of Dan.

I am wroth with all amongst ye save this damsel
    and aged man.

Save for these I had not spoken. Avoid ye the
    she-wolf's lair.

Of the hill of the great Dayfather I say unto you,
    beware.

If your course be west, sail westward ; whither, I
    would not know,
For the door of Janus is wide where'er I have
    will to go.
If I find ye, be ye heedful. My sword blade is
    short and strong,
And my shield as a wall before me. Bind me
    not with a thong,
Lest wolves in pack be upon ye. Julus hath
    many mates
That snarl in the lair, but howl as one from the
    towers and gates."

(2) The Servant of God stood silent, and gazed in
    that strong man's face
With eyes like starfilled sapphires as he spake of
    his name and place,
Then bade his thongs be severed, that each
    before each might stand
Eye upon eye ; and we parted ourselves upon
    either hand
As the prophet lifted his gaze to call down bless-
    ing and curse

Unto kindreds and peoples and times, unto
  better hap and to worse,
Whilst that chief stood silent, proud, in his eye
  the forward gleam
Of a shield on a wall that holdeth the sun with a
  steadfast beam.
" Thou art set in the night to watch. The towers
  of thy watch are seven.
As a strong man armed thou shootest thine
  arrows at highest heaven.
Did not I see thee afar by the Bozrah with long-
  built walls.
Thou bendest three spears beneath it, upon the
  latest it falls.
Thy swords are many and strong, thy quiver is
  wide and full,
Thy shafts are swiftly sped o'er all the plain of
  the bull.
Javan and Chittim are pierced ; Eber and Phut
  are low ;
Lud and Aram are stricken before the strength of
  thy bow.
Misraim is thine, and the half of Gomer's bands,
  and the Gaal.

All shall be given thy prey because thou hast
    cast down Baal,
On the silver wall of the islands thy farthest hunt-
    ing shall be
Ere the packs of the wolf are stayed by the
    dams of the stormy sea.
War is thy birthright, war is thy joy, and warfare
    thy bane.
Peace shall be very near thee, and under thee
    Peace be slain
In the street of the Holy City. Iron and brass
    and clay
Thou standest, and shalt be broken, thy watch-
    towers be for a prey
To the beasts of the field, and the fish of the sea,
    and the fowls of the air.
Thine helm is parted asunder, the crown of thy
    head left bare
To the winds of the East and the North. Out of
    Magog, Gomer, and Tur
With biting hail thou art driven, thy sword blade
    hath lost its spur
In the lap of thy wives, in the fulness of feasts,
    in the slavehood of power,

In thy fetters of gold thou art lost; yet there
  cometh a later hour
When swordless thou risest again with a woman's
  cunning device
Of tongue and snares of the eye the souls of men
  to entice.
By the Name thou hatest at heart, thou callest
  the nations afar.
The words in thy mouth are honey, but as worm-
  wood thine actions are.
This also long will I bear till the goats be set
  from the sheep,
For I set thee a watch of the night, and this My
  watch shalt thou keep."

(3) These things he spake to Julus and bade him
  hide in his heart
The blessing and cursing mingled, and gave him
  grace to depart
Ere we sailed betwixt mighty islands, both kept
  of a savage folk,
Now the Southward sells sons unto Egypt, but
  the Northerners brook no yoke.

Here the Prophet foretold how in latter days an
eagle should fly

From his eyrie amongst the mountains which
lifted heads to the sky,

Swift at the swarming of Gower, but lacking
strength to endure.

Unstable, his beak be dipped in the prey with a
hold unsure.

CHAPTER VI

(1) *A prophecy upon Eber ;* (2) *the ship cometh unto the Pen
of the Cape, and to Caer Melcarth ;* (3) *Elier the son of
Ziza greeteth its coming;* (4) *Tephi landeth and blesses
Elier and his sons, and is greeted by the Rock of the
Gate.*

(1) In short space we draw unto Eber, a land of
mountain and vale.

Purple and gold were its hills, and the Prophet
took up his tale.

"Thou art servant to Baal, oh Heber ; a servant
of him that shall slay

The leopard of Baal and his bull. Thy strength
  is taken away
Before the wind of the North, before the wind
  of the South
Till Gad and Tarshish arise to rend the bits from
  their mouth.
Swift upon wheels they roam, yea, wheeling,
  follow the course
Of the sun in his fields afar. They are each as a
  swift red horse
Wanton therein for a while. In their hearts is
  an evil thought
Lusting for things set apart, how low shall their
  lust be brought.
They are halt in their northward leap to the
  whitewalled tower of the sea,
Its warders shall overtake them, and great shall
  their burden be."

(2) Then drifting in calms to southward, we drew
  towards the Pen of the cape
Of the rock that keepeth the seagate and weareth
  a lion's shape

And watcheth both Phut and Eber, and inward
keepeth the sea,
And outward the endless waters that storm it
eternally.
A kingly strength it arises hoary and huge, the
crown
Of the pilot's hope who gazes. Thither the ships
go down
And may not avoid the watchmen. Narrow the
sea-gates are,
And Javan and Tursis stand where Canaan
holdeth the bar.
Their chapmen must chafer hardly with those
from the outer deep
For ivory, apes and gold and tin, with grain and
wool of the sheep ;
For Canaan found her pathways to the hiding
of men's desire,
And the spoils of all outer peoples have builded
the fanes of Tyre
Which shall fall, even now are falling. The
daughter of Zidon is low,
Is her burden not recorded, her nakedness,
shame, and woe ?

c

(3) Yet here was her mother her bondslave, cleansing
     her gate of the West

'Neath the Pen of the foot of Eber, and receiving
     therein her guest ;

For a strong Caer Zidon builded, and called it
     by Melcarth's name,

And Gad and Simeon were with her when into
     that cape we came

Under Elier the son of Ziza, who had knelt at
     Melcarth's shrine,

But was circumcised in his fathers, and cursed
     not the name Divine,

And knew the teaching of Moses, and ruled by
     the Book of the Law,

And yearned unto Jacob and David and that
     which their souls foresaw.

Six months he had mourned for Zion, but now
     in the seventh moon

He wept by the wall of his Caer from the dawning
     of day till noon.

His youth had been bloody and headstrong.
     His age was silent and wise.

And the men of Zidon obeyed him, and great he
     was in their eyes.

Now at noon he prayed unto Zion, and far on
    the eastern sky
Rose our sail. Then the son of Ziza cried with
    a joyful cry,
For the spirit of God was with him, "Prepare
    we a feast this day.
Six months was my fast appointed, but now it is
    lifted away.
My ashes are cleansed, pour forth a precious oil
    for mine head.
Set jewels upon my fingers to greet one sent by
    the dead.
My purple cloak shall be on me, my gems upon
    either ear,
My bracelets of gold, my breastplate of gold are
    meet to appear
In the eyes of those that bring tidings. Yea,
    yonder behold the wings
Of a dove, the branch in whose mouth was
    planted of mighty kings,
And watered of blood, and pruned that hence-
    forward it send forth shoots
Till its crown be lifted to heaven and earth be
    filled with its roots."

(4) Three hours ere set of the sun we came to the
strong-built wall,

Then the Prophet of God cried forth, and Elier
came at his call,

And knelt on the ground and answered of all
that he had prepared,

How his heart had leaped within him, and now
as a wand lay bared

And stript in our sight; and his sons knelt by
him on either hand.

That the Man of God might bless them as he set
his feet on their land.

But he craved my blessing also, that captain
hoary and grim,

So I set my palm to his forehead, and cried on
the name of Him

Who had chosen me out of Jesse, and lifted me
from the grave,

And out of the house of Pharaoh, and led me
upon the wave,

For a blessing on this man also, his sons, and
his strong-built town.

"Hail," I said, "to the rock that shall never be
overthrown

By the sea, but shall stand its warder, a keeper
    of many ways
To guard the treasures of ocean; and unto this
    town be praise.
Though its name be abomination, yet here is a
    shelter found,
And space for our feet to tread on that weary
    long for the ground,
And welcome of tongues that are near our own,
    and an open heart
To hear the cause of our coming, and bless us
    ere we depart.
Upon Elier God send blessing! Yea, as a lofty
    tree
Be his fourscore years an hundred to hold the
    Gate of the Sea.
His sons are many beside him. I bless them
    now, that they know
That when floods arise, the mountains are open
    wherein to go,
And hide and issue for prey or vengeance in
    flood or field.
They shall plough them both in the Springtime,
    and both shall a harvest yield.

This is the blessing of Tephi." Then he and his
sons arose

And cried my name, but their lips spake strangely,
and might not close

On its sound, for "Teia, Teia, Teia," these
Gaddites cried,

And "Teia, Teia, Teia," the voice of their rock
replied.

CHAPTER VII

(1) *The Canaanites set Melcarth upon their wall, but in strife
he is broken; (2) Hanmel calleth that place Carteia, and the
Prophet foretelleth the glory thereof; (3) The ship of Tyre
returning is lost with them that were in it; (4) Baruch
dieth at Carteia and a grandson is born unto Elier, and
called by his name.*

(1) Now some that bore Melcarth tarried to carry
him round about,

And high by his wall they set him, and named
his name with a shout,

But the voice of the rock replied not, for their
crying was shrill and small.

Then Simon the son of Elier shook his spear
at the wall,

A sign for the keeping of silence; and some
    that stood by the shrine

And looked for an omen, knowing the voice of
    their rock was mine,

Strove with the priests until Melcarth falling
    was broken in twain,

The image which Canaan brought from the
    uttermost eastern main,

And sent forth again to be with her sons, the
    toilers in ships,

That the name of their God might endure and
    be spoken of many lips.

A cubit he was in stature, and shapeless unto
    the crown

Of his head, but arms beside him in the likeness
    of man hung down.

In his right hand a golden trident was set for
    the rule of the sea,

And Elier bade it be plucked away, and be given
    to me.

(2) Then said he, " No more Caer Melcarth, Caer
    Teia this place is named.

Our rock hath shouted thy name.   Therein shall
 its walls be famed,
Whilst the seed of David endures," but the
 Prophet answered him, " Nay,
This too shall be broken in pieces, its stones be
 carried away.
Not once nor twice shall this be, by the land,
 by the seas, by the strait
Shall the spoilers come with engines to storm
 the tower of the gate ;
But at eve returneth a damsel that holdeth the
 twin-forked spear ;
A flaming wheel is her buckler, on all the isles is
 her fear,
And my daughter's sons are with her.   Hail
 to the thunder and smoke
Of the ships which vanquish the thunder, of
 her oxen brought to the yoke
To plough her by sea and by land a field for
 the harvests of peace.
From islands of iron she goeth to gather the
 world's increase.
Yea, islands of strength are the wheels of her
 chariot, her steeds shall not tire,

The storm is silent before them, their neighing
    is hailstones and fire.
Her peace is with winds and waters and clouds
    to lead her alone
Over every ocean wherein the might of her
    trident is known.
To the hill-top of hope, to the Holy Hill. Weep,
    weep no more
When the daughter of Zion sits in the gate.
    From the shore to the shore
Her name is heard in the echoing rock, her
    voice in the cave.
Her young lions draw to her side, though the
    fowls of the mountains rave.
Where the eagles gather together, is a lion in
    the narrow way.
He herdeth the kine before me, and setteth
    himself at bay,
If at dawn the eagles hover, and the prey that
    is in their beak
Causeth their wings to tarry, their eyries be
    far to seek
When the lion springeth upon them. Not yet,
    oh my daughter, not yet

Is thy seat on the lion of the gate, but let not
    thy soul forget."

(3) Three months beside the rock we abode, and
    here it befell

That the seamen of Tyre returned, and we
    knew how this hap was well,

For they murmured at Melcarth's fall, and
    therefore an evil thought

Arose in their hearts to slay us; taking that
    which we brought

Out of Egypt, the jewels of Pharaoh, which
    Sebnet his servant gave

When Pharaoh named me his daughter. These
    we cared not to save,

But the things of the Lord were precious. These
    things a slave, with a kiss,

Drew from the lips of a seaman, and Elier
    heard of this

And brought our goods from the vessel, and
    bade its sailors begone,

Though the Prophet told it to him how that
    ship should sink as a stone.

(4) Which thing hath surely happened, for at the
next eventide

When Baruch the scribe sat with us, his eyelids
were opened wide,

And he said, "The Lord stands by me. My
spirit is in His hand,

He slayeth Tyre in deep waters. He saveth
me by the land,

He holdeth me in dark places." And then he
tottered and fell,

And went to the house of our fathers with
David my sire to dwell,

Moses and Jacob with them; an old man
withered and hoar,

Whose eyes wept blood over Zion, the tale of
his years fourscore.

We buried him by Caer Teia, and there in the
lisping tongue

Of its folk men prayed above him, whilst songs
of the grave were sung

By me and my women duly. On that same
night at the morn

To the wife of Simon Ben Elier a fair man-child
was born,

And they named him Baruch from him, This
child is amongst you still.

Simon Breach ye name him. In our speech
this is good and ill,

As of one that is striped and spotted, but fierce
though his angers be

His name shall be known unto after days for
his faith to me.

Chapter VIII

(1) *A ship is sought of Necbal, a Canaanite of Dor, who captures
a ship of the Greeks; (2) A prophecy upon the body of
Aias; (3) Nabal prepares the ship which cometh without
hap unto Tarshish.*

(1) Two months we took much counsel to find us a
further aid

For our journey beyond the sea-porch, but at last
a pact was made

With a Raclen who came out of Lud, but in
Canaan, Dor was his birth,

And he traded in many waters to all the ends of
the earth.

Aine, a daughter of Dan was the mother of
    Necbal. She knew
Where Dan lay coiled as a serpent; watching
    all birds as they flew,
Naming those that passed to Eriu when winter
    was over and spent.
She also had sorrow for Zion, her locks and her
    garments were rent,
But she joyed in the surety of Dan, his salvation
    sealed of the deep,
Where in grasses and long green rushes the
    broods of the serpent creep
To sting the horse with its rider, the ox and the
    lion and lamb,
Until all be gathered together in the promise of
    Abraham.
She aided us much with Necbal, who bade his
    miledhs await
A ship of the isles of Elissa that sought to steer
    by the Gate,
For the Fœni brook no rivals to hamper their
    raclen's mart,
The weaker come not anigh, the stronger they
    bid depart,

Or fight for the way in the narrow porch, so the
    miledhs of Dor

And of Rhodan took that ship of the Greeks, and
    to Necbal's store

Her riches were brought, yet brave and fierce
    were the men of her crew

Ere Achæas and Aias her captains the bands of
    the Fœni slew.

I beheld the body of Aias, a mighty hero and
    strong,

His spear was stayed to his wrist by its plaitings
    of leather thong.

His greaves were of brass, and his helm was
    brass, and his full-moon shield

Was pictured with tales of his sires that had
    harried the Ilian field,

Chiefs of the Raclen, and princes of Dan in his
    islands, and lords

Of the men of Argol and Chittim, and captains
    that went by the fords

To the parts about Inis Colcha for fleeces and
    golden dust,

And fair-haired bondslaves whose fathers will sell
    their daughters to lust.

Thou wast shapely in death brave Aias, and crisp
    the curls of thine head.

Thy feet and thy hands were little, yet thine arm
    was mighty to shed

The blood that had drenched thy sword-blade
    when thou heldest thy ship alone,

Till caught in the nets of the miledhs at last thou
    wast overthrown.

(2) Now the Prophet said " Out of Javan and Tiras
    a ram shall rise,

To storm the gates of the sun in the golden house
    of the skies.

Even now is born God's servant to Madai.   Him
    shall He bless

To the height of a moon whose splendour shall
    weaken yet not be less.

By him shall Bel be broken, with Misraim, Lud
    and Tyre,

But the ram of Elissa, the two-horned ram, shall
    tread him in mire.

He breaketh the walls before him, he butteth the
    furthest East.

The Holy Hill shall know him.   He setteth foot
  on the beast,
Upon Egypt, o'er Elam and Assur he goeth
  abroad at will.
The Bamah beholds his horsemen.   The roof of
  the world sits still.
He is feared in Hinda and Ganga, and on to
  their utmost isle
As none that hath come before him.   Yet, behold, in a moment's while
He is ended and gone, the place of his ending
  holds not his fame,
But the place of his rest shall be famous, and
  ever dwell in his name.
The wise shall write him in story, the cunning
  picture his deed.
His pride is a garnered treasure whereon shall
  the ages feed.
Magog and Gog adore him.   Shushan claimeth
  his right,
But the ram of Helle is set in the sky as her
  beacon light."

(3) Now Necbal plundered the corpses, and lent us
    the strong-built boat,

Building great stones within her that upright and
    strong she should float,

For rowers we had not as yet, and trusted but to
    the sail

To lead, and the stones within to steady us unto
    the gale,

If the winds should beat upon us, and wild sea-
    horses outcurl

Their manes on the plain, but Gaddites and
    Fomorcs we had to furl

Our sail in such hap from Elier, who, blessing us,
    bade farewell,

Sending Simon his son with his babe and wife
    to guide us and tell

The shallows, and count the headlands as we
    sought from the western bay

Of the Gate, north-west by the sun, where the
    island of turrets lay,

Near the mines of bright iron and copper, and
    the wind of the south-west still

Blew soft on our sail, so thither no hap of our
    voyage was ill.

D

Chapter IX

(1) *Ith the Prince of Breogan giveth rowers to the ship ;* (2) *He maketh a song for their guidance ;* (3) *Ith speaketh of his son Lugaid ;* (4) *Tephi parting from Ith, the galley is brought by a storm unto Pen Sauel.*

(1) Now we came unto Ith to Tarshish, a miledh
   of war was he,
   A fierce sea king that ever had joyed in the
   stormy sea,
   The crash of the prows in battle, and coast
   towns given to flame ;
   But for Elier's sake he loved us when unto his
   courts we came.
   He gave us slaves of the Nemidh, lusty, freckled
   and strong,
   To fill the bank of the oarsmen, and bend their
   backs unto song ;
   And he made them a song to swing to as onward
   we went our way,
   And I wrote that song before him, and helped
   them to learn its sway.

(2) To the star, to the star, to the star, do we row
   At the eve, in the dawn, through the day,

Seven moons, seven nights do we sit as we go
  By the coast of the hills on our way.
To the East, to the right, sixty hours swing the oars
  To the cape of the fire-bearing Pen,
From its tower is our travail to come by the
    shores
  Whereon Net of the Stones hath his den.
We are swift, we are strong, for the seas are alone,
  And the hills of the wave builded high,
And the sea god hath made him a place for a
    throne,
  And the Thunder his camp in the sky.
By the cahirs of Net, by the stones which he built
  Are the streams where our weary may drink.
If his men give us hurt unto Ith is their guilt,
  And their names in his nostrils shall stink.
To the West, to the North, to the East by the
    heads,
  Out of Caerned count forty and four
Till our way goeth north by the coast where it
    leads
  Past the woods of the wolf and the boar.
Wait the sun lest the sea-witch draw cloud to her
    hand,

With the moon on our stern must we row,
  Whilst the eyes of the watchmen await on a land
    As a blue mist, as blood or as snow.
  He is blue where he watches the storehouse of tin.
    If his beard we may pluck, he shall smile,
  To the house of the bond-slaves of Ith we go in
    To Elatha, and rest us awhile.

(3) Now Ith regarded my singing, and grace in his
      eyes I found,
    And he said, "I have mourned my son, who has
      fled beyond the bound
    Of Eber and Gad and Breogan, perchance he
      hath passed away,
    But I would that Lugaid were with me, and thou
      wert his bride this day.
    My sons are not few, but Lugaid's mate should
      be far to seek,
    He was first in arms and in leechcraft, first in
      the stithy's reek,
    First in counsel or pastime, and first would he
      be in pride,
    So he brooked no king above him, and forth he
      went from my side.

Yet my heart is weary for him, and never hath
    yearned again
As it yearneth to thee my daughter; and glad I
    were if the twain
Could meet if indeed he liveth. Thou art little,
    but thou art wise,
Thy words unto men are few, but queenly their
    message lies
In the hearts of slaves thereafter. Now, there-
    fore my daughter plead
With my son where'er he greets thee, and his
    ears shall give good heed."

(4) Now the Nemidh and Fomorcs sang, setting
    their backs to the oar
Many days till they swung together, and the
    chief of the rowers swore
That with such he feared no evil. So we went
    from the fortress of Ith
Well stored with garments and trinkets, and
    many a gift therewith,
Brooches, armlets and rings in caskets of ivory,
With mirrors of bronze and combs of shells of
    beasts of the sea ;

For the hand of Ith was open, if wide, uncomely
and red,

And he loved the message of Elier, whilst Simon
his son had wed

His nigh of kin, who remained with her husband
behind when we went;

And I gave unto Ith three gems to witness my
soul's content,

Blue, green and tawny, of Egypt; and the
Prophet said, " Let the blue

That is alway before thee lead thee to seek the
gift that is new.

Lo, the mine of emeralds is deep. This, therefore, shall be thy seal

Of a mining far in the deep in green forests of
Ar Brazeel.

In the tawny stone, behold it, thy path is set to
the South,

And the tawny sands poured seawards from
many a river's mouth.

Thy wealth is in this, in the yellow sands, in the
shipmen's trade,

In the tawny lands there is none to make thy
Breogans afraid."

So spake he to Ith at our parting, and sad are our
　　hearts to go
By the side of the deep-hued hills, whilst the
　　Fomorcs and Nemidh row
To their song, but the sea song cheers us ; and so
　　we pass without hap
To the Firepen flaming northwards that watcheth
　　on Eber's cap.
There, casting the Pen behind us, we flee for the
　　north in fear,
For the sea-snakes coil beneath us until we may
　　hardly steer,
And our galley is tossed up endwise, and some
　　of our oars are broke,
And some break hearts of our Nemidh, and white
　　are my womenfolk ;
But I sing them the psalms of David, and how
　　he escaped of Saul
When the Lord his God stood by him ; and
　　raised his feet on the wall
When the might of man availed not.　Whilst the
　　Prophet readeth his scroll
And recketh not of the stormwind, nor heedeth
　　the water's roll,

For the Word of the Lord is in him.   In a noon
    that is black like night
He beholdeth the heavens open.   His face is a
    shining light.
Then Buchi breaketh the pole of the helm, and
    we may not steer,
And he clings to the mast beside us, and heareth
    our holy cheer
As we go unholpen of man ; but the mighty hand
    of the Lord
Is with us, and far before us the signs of his
    grace outpoured.
The seamen's marks have failed in the storm, and
    the watchmen dream
We are lost in plains of the ocean where never
    the seabirds scream,
And no life save of sea beasts liveth ; but Buchi,
    the wise man, told
Of one who had sought Ar Brazeel, and its city
    whose towers are gold,
And came on that island westward, and stored
    his ship and returned,
And after six months found Tarshish, a bearer of
    thoughts that burned

In his bosom whilst he hid them ; for a pestilence
    found his crew

And strewed their bones upon ocean, and all save
    himself it slew ;

Whilst himself died little after, leaving with Buchi
    his thought.

Therefore Buchi enquired upon us if now that
    island be sought,

When our oars were mended and manned, but
    the Servant of God forbade,

And counted us yet four days wherein our souls
    should be sad.

Commending us prayer and fasting.   Then, there
    fore by night I prayed,

And by day I heartened my women in God, and
    was not afraid.

Now, storm was yet on the fifth day but lessened,
    and looking forth

In the cloud methought that there gathered a
    darker cloud from the North,

And enquired of the son of Helek, who shaped as
    an arch his hand,

And gazing, gave thanks unto Heaven that
    brought us in sight of land.

Then we saw it as isles and a wrathful cape, for
    ragged and grey
The rocks ran down to the sea, and shewed us
    no entrance way.
Whilst our helm was broke, but the Lord of the
    sky commanded the wind
To save us out of their teeth in a haven that lay
    behind,
Where a Pen arose to the East, and a marvel of
    God in that Pen,
For the storehouse of Ith stood there, and the
    place of Elatha's men.
More swift than by any road that our pilot had
    steered,
To the land of tin were we come, yea, even unto
    his beard.

CHAPTER X

(1) *Elatha and the servants of Ith give welcome at Pen Sauel ;*
(2) Elatha sendeth men to Eriu for tidings ; (3) They
are sent back with gifts unto Tephi from the men of
Eriu, and a welcome thither.

(1) BLESSED were we in the Lord when the traders of
    Ith came out,

And learning our message towards them, raised
    his name with a shout

And brought us into their houses beneath the
    Pen of the wood,

Slaying an ox and seething its flesh in pots for our
    food,

And baking fish with corn and herbs that grew in
    their garth

Beneath the strong steep Pen whereon was
    builded a rath,

Defender of lead and tin, and black stones out
    of their mines,

Both that which burneth as wood, and that
    which glitters and shines

Betwixt the breasts of their damsels. To the
    mines were our Nemidh sent

To toil three years for their master, nor thus
    were they ill content,

For we gave them a promise from Ith, that
    after three years should come

A ship out of Kirjath Hadtho, and bear them
    unto their home

Where the eye of day is clear on the rocks
    without cloud to blind,

And the dates are sweet in the mouth where the
bowman seeketh the hind.

(2) Then Elatha the kinsman of Ith gave counsel to
rest awhile
Till swift boats be sent to Eriu to question the
men of that isle
Where the princes of Dan abode, and chiefly o
Jochad, the son
Of Duach, him that their landsmen had chosen
as Heremon,*
Whose fathers came out of Japho wherein they
were held too straight
By the kings of Gath and of Eckron, and spreading
their sails to fate
Drew their swords unto kingship in Chittim,
Rhodan, and Lud,
And ruled Ar Kadesh, and mingled the stream
of the chosen blood
In many a mountain torrent, on many a peopled
coast
Ere they lighted on green Eriu a little, a noble
host,

* Ir. Eocaidh.   Gr. Achaios.

Which fought the cause of the landsmen. This
    fame, and their names herein

The Prophet foreknew of the Dannites, the
    furthest of Jacob's kin.

With these he would leave on the sun's path the
    twig of the lofty tree,

The small green bough of the olive, in the midst
    of the deep to be

Even yet in Abraham's bosom, the home of his
    sons afar

Who replenish their strength in the isles, ere
    they gather to seek the star

Of Isaac and Jacob their fathers, when Israel
    filleth the earth

With joy in the sound of his coming, and music
    and songs of mirth.

(3) Five weeks we abode at Pensauel till the men of
    the land returned

With tidings whereat the Prophet rejoiced, and
    my spirit burned.

At ⌊Pen Edair they heard of peace, how Eriu
    yearned for the choice

Of a guard against evil rulers, and the *aire* *
  cried with one voice

Upon Jochad, the son of Duach, a prince of the
  tribe of Dan,

A champion wise and mighty, and sprung of that
  chosen clan

Which had captained miledhs in Javan, and
  their hosts throughout Eberled.

This prince had been sought for of many, yet
  stayed in his prime unwed,

For the ollamhs that watched the stars to the
  twilight whereon he was born

Beheld ere the sun's arising a moon with a
  slender horn

Ascend from the sea before him, to lead his
  light out of sleep ;

And they set on the babe a vow that the
  strength of the man should keep,

To hold himself from the stars, till a moon in
  the eastern sky

Should shine in the dark and lead him, yea,
  even when noon was high.

* *Aire*, yeomen, literally ploughmen.

For that moon abided near him till over him
    clouds were grey,

And at eventide was seen ere the sun was
    hidden away.

Now there went by the men of Elatha as a token
    to Pen Edair

The slender horns of silver, the clasp I was wont
    to wear

On my veil in the house of my fathers. The
    daughters of kings were known

By such from old days before me, and my sire
    upon David's throne

Had fastened the clasp upon me, when they
    brought me first to his sight,

Though " Tephi " * he cried in anger, and in me
    had little delight.

This token the Prophet bade me loose from the
    folds of my veil

And send as the horns wherewith he should
    harry the priests of Baal ;

For he sent a fiery message forth by Elatha's men

Who told it the chiefs of Eriu, and they that
    dwelt by the Pen

* A small one. Mignon.

Of Edair scoffed at its hearing, taking the tale
    for a jest

To be told in the near assembly where the war
    chiefs gathered at rest,

But when Jochad the Prince had heard it, he
    straightway rose from his seat

And cried, " It is twilight still, but the day shall
    be soon complete.

Ye have doubted the dawn, ye chiefs of Canaan,
    Eber and Finn,

But the moon on the furthest deeps hath reached
    the island of tin

To shine full soon o'er Pen Edair. Her shadow
    cometh before.

At her rising the fomorcs * shall flee and the men
    of Eriu adore.

Bring in these men out of Albion, and bid the
    ollamhs unroll

The message they bring with the token from him
    that hath writ us a scroll.

Then were called the men of Elatha, and unto
    the warrior's hill

* Sea rovers.

They came with the scroll of the Prophet, and
    none spoke kindly or ill
Whilst Sri the son of Ezru, an ollamh skilled in
    the speech
Of Zion, Nemidh and Breogan, held forth his
    hand unto each
And took from the one my token, and bowed to
    the Holy Name
On the Prophet's scroll, and sought it of his fellow
    that with him came,
And read its words in men's ears.   Great was
    the import thereof,
For the Lord had spoken therein.   Now the last
    of His word was love,
But wrath was in the beginning, which the chiefs
    waxed wrathful to hear,
And murmurs arose in their midst both of anger
    and scorn and fear.
" Ho, ye that dwell in the rushes,—Ho, ye that
    walk by the sea,
Afar, in the clear-walled island,—Ye have whored
    and are sundered from Me.
Ye are set upon idols greatly,—Your feet are
    clayed in the mire,

E

Ye are fat with the flesh forbidden,—Your foreheads swell with desire.

As swine ye rush on each other,—Ye gore as an
unclean beast.

Your prayers are evil before Me,—My soul ab-
horreth your feast.

Ye are long cast out from Zion.—Your feet were
the first to flee.

Ye have spawned in Javan and Nimidh,—Your
seed is lost in the sea.

Jacob is wasted in Eber,—Yea, as a wine that is
spilt.

The poison of asps is in you,—Have I not known
your guilt?

The glory of Zion was yours,—Ye first have
hastened her fall.

Weep for your sins, ye faithless,—Weep not My
Temple's wall.

For now I dwell not in houses,—Only with men
I dwell.

Hearken now to My message,—Hear it and heed
it well.

I call and ye shall not hearken.—I cry, and ye
will not heed.

The blessing of Abraham liveth.—I sow you with
    David's seed.

A little seed unto ages.—Ye shall tread it under
    your feet.

It shall sleep amidst your tumults.—It shall
    slumber in cold and heat.

My burden on Eriu is broken against you, the
    thing I crave

Is a name forgot, and a secret place, and a far-
    off grave.

My name I have left in Egypt.   Unto an hiding
    place

I bring the treasures of Yahveh that He shutteth
    from every face

Till this season.   Not unto Dan are these, but I
    bring therewith

The daughter of David, daughter of Pharaoh,
    daughter of Ith,

A fount that Yahveh hath cleansed, anointed of
    Him from birth,

Heiress of tribes and peoples scattered o'er all
    the earth.

The furthest isles are her portion, the sea is hers
    as her dower.

Her sons shall rule in Eriu, her sons' sons reign
    unto power ;

Till her child that shall be, gather the flock of
    David anew.

His head is crowned with the sun. His feet are
    wet with the dew

As he leadeth them in the morning. This also
    ye may not learn :

Ye are blind, but a ring in the snout, is plain that
    ye all discern.

Behold her silvern crescent which marketh the
    daughter of kings.

A king that wrought evil gave it. Moreover,
    bracelets and rings

Be hers of Tarshish from Ith of the Breogan out
    of his hold

Wherein ye barter your herds and harvest for
    treasures of gold.

He is greater than ye, yet the seed of Judah hath
    known a sire

Higher than Ith, for Misraim bows to its Lord's
    desire ;

And he gave to his daughter Tephi royal garments
    that shine

As sunset, and are as the rainbow with jewels out
of the mine.

Who is he that sitteth amongst you shall raise his
eyes to their hem.

The Queen of the Gates and Nile cometh out of
Jerusalem

As a sweet fruit ripened in Winter. Hither with
her the Stone,—

The Stone of the Kingdom cometh. It shall not
be left alone

Henceforth of her sons for ever. I bid ye prepare
her a home

Wherein all shall be meet and ready that the feet
of the Queen shall come,

Yet not ask of me. I am left in Egypt a pillar
to be

Unto days and lands and peoples, when the Lord
bears witness in me.

I stand a sower, a ploughman. My God hath set
me to plant.

I shall not fail in His time. His hand hath
holpen my want.

A builder, I set one stone ; as a husbandman, a
seed ;

But the Stone is the dwelling of Him from whose
    hand shall the nations feed,
And thereon shall rest His Chosen, whose King-
    dom is East and West,
Whereupon the sun shall wander and find no
    place for his rest
Of the night, but day endureth.   Heed ye this
    work, and mark,
At the end of days it is clear.   It is dim in the
    veils of the Ark.
This also may not be broken, though men shall
    hide it away,
It standeth in earth for ever, and ruleth the night
    and day.''
These things read Sri in their hearing, and silence
    dwelt for a space.
The hearts of the warriors held them, and each
    man sat in his place
A dreamer of far-off places, and pondered on
    hidden things,
And thrones and kindreds and seasons and sons
    that should reign as kings ;
But the children of Baal were angered, and
    Tuirbhi was first to speak,

The chief of the Tyrian craftsmen. "What came
    ye hither to seek,
Ye men of Elatha, the scourge of the fomorcs, the
    shipman of Dan,
And foster-father to Jochad? I know the wiles
    that ye plan!
Elatha's mines are empty. His smelters handle the
    spear.
His sails are gathered together that Eriu may
    dwell in fear.
Ye are come as spies before him. Answer ye to
    his boast,
That the men of Eriu be gathered to greet him
    on every coast,
Though Ith out of Eber help him, and Elier out
    of the Gate.
If Egypt indeed be with him, it is long that his
    host must wait.
But come ye many or few our firbolgs have little
    fear
Whilst Tuirbhi watcheth his anvils to furnish each
    with a spear.
By Caiseal the stones are strong that are piled
    upon Breogan's wall,

And the crag of Edair is steep whereupon it is ill
to fall.

Our gold is stored in the mosses, our oxen hidden
away,

Are ill to hunt in the mountains, and few shall be
for a prey.

Though he send the chief of his Milidh, surely
we will not stir ;

Though he send his champion to Jochad, ill shall
it be with Ir.

For Ir, his captain of strength, the wild boar
rooteth a grave.

If he come to the land of Eriu, his ships shall
burn on the wave

Though Jochad his brother help him." Thus
Tuirbhi spake and was still,

And Elatha's men stood silent, nor answered they
good or ill.

But the bard of Jochad endured not. Ethan,
Muiroideach's son,

A youth, but a mighty singer that ever the oak-
wreath won.

In wrath he arose, and sang against Tuirbhi a
song of might

Till his brow set red in his bosom and his heart
    was closed from the light.

" Hear ye, hear ye, ye princes.—Hear ye, the son
    of the smith.

Stand in the blast of the bellows,—Be ye all
    shaken therewith.

Give your nose to the pincers,—So doth he
    lengthen it out.

Crafty the rings of Tuirbhi,—Gaily they hang in
    the snout.

Bowed in the back is Tuirbhi.—Are ye not all
    the weight?

Doth not he squeal beneath it?—Doth not a
    beldame prate?

She is blind beneath her forelocks.—Is she not
    sore afraid?

Shall Ir at his coming take her?—Shall he choose
    the smith as a maid?

Let laughter be upon Tuirbhi,—Go clothe his
    brawn with a smock.

Clip his bristles to smoothness,—lest the men of
    Elatha mock.

Those that have brought good tidings,—See in
    the hand of Sri.

A slender silvern crescent.—The moon of the
    East is nigh.

Her horns are peace and riches.—Set as an elfin
    queen

She saileth her boat in heaven.—Her rounded
    fulness hath been

Before and it shall be after.—She hideth yet for a
    space

From Eriu in her chamber,—He findeth her
    hidden place

He rejoiceth in her beauty.—Robe Eriu like a
    king.

Set purple and gold upon him.—May a sun arise
    to fling

His mantle of gold about her,—his fires in her
    slender form,

That her months be duly rounded,—That new
    stars in the sky be born.

She hath gems to teach the springtime,—veils to
    shelter the heat.

Gold for the Autumn harvest,—Her light in
    Winter is sweet,

Fair on the snow she glistens.—We dream of that
    which may be.

Our hearts are where she riseth,—In isles of the
    Eastern sea,

In mighty cities and temples,—in stories of
    ancient days,

In visions of kings and heroes,—with priests
    amidst songs of praise.

Go forth to meet her, my soul.—My belovèd is
    very fair ;

She is white, she hath eyes as stars,—The night
    is set in her hair ;

She hath rainbows in all her garments,—She hath
    dewdrops about her throat.

Her hands are slender lilies,—Her voice hath the
    cushat's note.

Her lips are as winter berries,—Her foot hath a
    mouse's fall.

Where she cometh joy awaketh,—He riseth to
    festival.

Three mighty kings are her sires—No king's son
    sits at her side.

She cometh a queen to Eriu,—A queen and a
    chosen bride,

Eriu shareth her birthright,—The flower of its
    greenest sod

Shall blossom here in our midst,—and grow to
the Land of God."

Then the chiefs of green Eriu rose up from their
seats to throng

To the place of Ethan, and raised him aloft and
bore him along

On a shield and shouted and crowned him, for
seldom such tongue was heard

As Ethan's, strong as a stormwind, clear as a
morning bird

Was his voice, and his touch on the harpstrings
light, like a fountain's play,

A ripple of running music that chimed with the
voice alway.

Oft have I heard, and loved it.   Ah me, that a
bard be slain

By the coward deed of a churl, for a witchwife
light and vain.

Each chief gave then a guerdon which matched
with the giver's state.

First Balor grandson of Net flung down twelve
pounds by weight

Red gold in torcs and armlets.   Heavy his herdsmen's toil.

Then Crimthann Lord of Pen Edair gave him an
    ocean spoil

Of goblets and horns of silver, and Nuadh of
    Usna's keep

Gave gold and horns of a seabeast brought from
    the northern deep,

And the chiefs of the merchants gave him a
    breastplate of well-wrought gold,

With an ivory chessplay carved by cunning men
    to the mould

Of kings with their chiefs and firbolgs. Such
    bard gift ne'er hath been gained

As Ethan's, a hundred warriors plucked their
    cloaks till it rained

A shower of their flashing brooches; but Jochad
    his lord came late,

Yet foremost, for Jochad was proud. His gold
    was little of weight,

He had not oppressed his yeomen, yet he gave
    unto Ethan's hand

A gift which was more than Balor's, and worth
    the half of his land,

A brooch of red gold which wizards of Tursis had
    sprinkled o'er

With a golden sand by magic, and out of their
     hidden lore

Had heaped it in flowers and bosses, and mar-
     vellous stems of fern

Where the eye was 'wildered in choice, and scarce
     had strength to discern ;

Yet the whole was a sun in glory.   Now, once
     that glory was seen

With Eileen fairest of women, she that was set as
     a queen

O'er Elissa in fair Ar Galish, and fled to a further
     shore

To carry the curse of Javan, and leave her tale
     evermore

In the mouths of bards and singers.   Now
     Jochad's sires out of Troy

Won this when the city had fallen, a treasure
     without alloy

In the eyes of all fair women, a spell compelling
     the eyes,

A gift beyond price more precious than aught
     that the merchant buys.

Then Ethan cried, " With a bardgift, lo, I am
     made a prince.

Such hansels may not be handled, mine eyes at
    their brightness wince.

Cover them all lest they blind me.   Let them be
    carried away.

Let these be earnest of Eriu that the moon no
    more shall delay,

But hasten her speedy rising."   Then the
    chieftains shouted loud

" Let us see the moon of the morning. The
    edge of whose silver cloud

Hath touched upon Albinn. Seek it. Ye men
    of Elatha speed

With the greetings of green Eriu to welcome the
    chosen seed

Of the Daogdœ, kings of Morias, that holy city of
    fate,

Morias Fail of our fathers. She mourneth its
    fallen state.

Both in Egypt she mourned, and in Breogan, but
    tell her that warm shall be

The hearth that is lit in Eriu, the greenest isle of
    the sea."

CHAPTER XI

(1) *The men return to Elatha, and give the gifts of Eriu unto*
   Tephi, who telleth of her chief jewels ; (2) Bres telleth
   his father of the prophecy upon Jochad ; (3) Elatha
   mourneth for their departure and communeth upon
   Ephraim with the Prophet ; (4) He prepareth many
   vessels for them, and sendeth Bres with them to Eriu ;
   (5) *of their journey thitherwards.*

(1) So these men came joyful to Albinn, and poured
     their tale in our ears,
   How their hearts were low at Pen Edair, and
     heavy at Crimthann's jeers,
   And sunken at Tuirbhi's boasting ; but how from
     the side of a chief
   Clearbrowed as the dawn sprung a youth who
     had given their souls relief,
   Heaping out wealth upon them.   Then they
     brought the bardgift they bore
   From the chiefs and Ethan, and showed it.
     Now behold, the first of their store
   Was the wondergift of Jochad.   Mine eyes grew
     blinded thereon
   And Elatha took, and laid it on my breast in
     place of the stone

Of Pharaoh, a sky of turquoise that swam betwixt
    golden wings,

A precious gift and an holy, and meet for
    daughters of kings,

Chosen of God and his Servant, for the Lord had
    shapen its thought

In its maker.   Where graven idols of beasts have
    made Him of naught.

His thought shall behold their ashes, and the
    wings of His spirit fly

Before men's souls in their blindness to name
    Him eternally.

So I changed the place of my jewels, my moon I
    set on my brow,

And the turquoise lay at my throat where it
    wideneth out below ;

But the sun of Eileen I planted deeply upon my
    breast.

There it shall gleam in my sidhe,* and lighten
    the gloom of my rest.

(2) Then Elatha spake with the Man of God, and
    called upon Bres,

---

* Tumulus with chamber at centre, pronounced *shee*.

His firstborn, the stay of his age, that now was
    his strength to press
The presses of Eriu and Albinn, and thrice had
    been unto Ith,
To Tarshish, and once to Caer Teia, and bade
    him unfold the myth
Of the bards upon Jochad's cradle, for the twain
    were nurtured as one.
When the father of Jochad fell, his babe and his
    only son
Shared couch and cover and breast with Bres in
    the fort of the horn
Of Albinn. So Bres well-skilled in that legend
    of mystic morn
Gave forth its tale in our hearing, and I treasured
    it in my heart,
Ere Elatha gathered his vessels and gave us speed
    to depart.

(3) Now Elatha communed much with the Prophet,
    and wept and grieved
Upon Zion greatly, but read the promise and
    greatly believed

The blessing of Jacob on Joseph and Judah,
  beholding the day
When Ephraim's kiss should bind them, and sin
  be taken away :
And he learned by his packmen where Ephraim
  tarried now by the path
Out of Hara, Haber and Halah, wherein the Lord
  in his wrath   .
Had set him amidst the Madai, and how by Kir
  he had fled
Through the children of Heth to the mountains,
  and crossed by the watershed
To the summer land Defroban, and built him a
  temple there,
For the Lord in the pastures of Kef, and now
  the name which they bare
Was Asirgard, City of God, that the God of
  Moses therein
Might keep him from Heth and Magog, and
  purge him away from his sin.
Now Elatha blessed the Lord beholding how
  David should wait
The kiss of Joseph whose ploughing tarrieth long
  in the gate.

The Engel is slow and heavy and loves by the
river mead

To lie in the sun by day, and rise at morning to
feed.

But hateth the yoke and the plough for the field
wherein she would lie

Where the lion is in the gate. Yet the Engel
shall draw anigh

For the ploughing, and harvest shall whiten slowly
up from the blade

When the boughs of the planted cedar are over
his head for shade.

(4) Of these things Elatha communed much with the
Prophet and bade

That the lioness cub of Judah be with such pomp
arrayed

As the power in his hand might furnish to pass
to that seagirt isle

Wherein is the sapling planted to suck the dews
for awhile,

Ere it grow of strength to return to the land of
the strong free breeze,

And increase on its northern mountains, and
    spread to its narrow seas.

By its shores of grey-blue granite, its shores of
    blood and of snow,

By all walls of its fertile garden fenced of the sea
    shall it grow.

Therefore he painted his vessels, and set them
    with snowy sails,

And bound green wreaths to their foreheads, and
    out of his merchant bales

Brought scarlet and blue and white to flutter
    upon the mast

And stripe their sterns with a rainbow to oaken
    planking made fast.

Then men of the silvery isle of Vect he chose
    for our band,

An island of many havens that lieth under that
    land ;

And mixed folk out of the Domnan that dwell
    where the tors are red,

Mighty men of the sea, fire-hearted, wary of
    head ;

And fisherfolk from the horn, the beard of the
    promised isle,

A mixed folk also whose maidens hark to the
raclen's wile,

'Till the blood of Zidon and Israel toileth amidst
the veins

Of the rocks wherefrom the princes of the Tyrians
suck their gains ;

And fomorcs * of Khumru north till then reachest
the furthest Pen

Of Lochlann, returning again by coasts of mountain and fen

To the narrow seas of Albinn by the shore of the
silver wall,

And pass by the island of Vect again to Elatha's
hall.

A hundred ships had Elatha, and he gathered
fifty and three

With chosen men as their pilots, to make a con-
voy for me,

And the wealth of Egypt and Tarshish and that
which Eriu gave,

That my sailing be spoken of many, my path be
sure on the wave,

* Fomorians, sea-rovers.

And Eriu have fear and joy at my coming. Two
    thousand and five

Were the living souls of our navy. " A gallant
    swarm for the hive

Of a queen well stored with honey." Thus Bres
    of the miledh spake ;

And his father answered again, right glad for his
    firstborn's sake

(The son that Delbaeth's daughter bare him in
    Maoth Seein

When she loved his youth ere she fled with the
    sea-king to be his queen)

" To thee be the hiving of her," and, Bres being
    merry, cried back,

" How may I store the honey with all the wasps
    in its track ?

Thou knowest our wasps of Eriu." Whereat
    Elatha replied,

" The Lord shall harbour the queen-bee. Be
    thou but found on His side

And His sweetness shall surely bless thee." Such
    answer more grave than gay

Had Bres from his father Elatha before we went
    on our way,

With the summer breeze behind us.   We jour-
        neyed first to the North

Beside the lands of the Khumru which deep in
        the sea jut forth,

Till we came to their holy island, and were blessed
        of their ancient bards

Who sang to their harps the night of our resting,
        but afterwards

With a clear east wind ere dawn we went by a
        path that lay

To the West, and brought us swiftly in sight of
        the fairest bay

Whereupon I had looked.   By Edair our anchors
        and stones we cast,

And the firbolgs of Crimthann swam with ropes
        to steady us fast ;

And Crimthann came with his captains and stood
        to watch on the strand

And shouted, and many bards sang welcomes of
        Eriu's land.

Chapter XII

(1) Tephi setteth foot upon Eriu, the defended island; (2) The
song of the bards at her landing.

(1) THEN looked I for Ethan and knew him, for his
    voice was sweetest of all ;

But his lord I might not know 'midst the chiefs
    out of Crimthann's hall,

Twelve warriors strong, but I liked not them-
    selves in their cloaks of red :

So I deemed the master of Ethan a dullard, and
    bowed my head,

And wrapped his sun in my mantle, ere smiling
    I raised one hand

To my women, whilst out of the ship I was
    carried in haste to land

By Ethan the bard, green-mantled ; and another
    that, clad as he,

Throwing his harp on the pebbles, ran singing
    still through the sea,

And raised up his arms imploring, till my women
    lifted me out

To the seat they made with their mantles.   Nor
    did I tremble or doubt

For their tread was steady and sure ; and I
smiled to him to the right,
For his brow was clear and steadfast, his eyes
were joyous and bright ;
And so by the bards of Eriu I was borne through
the shallow sea,
And this was beginning of joy and pain in the
heart of three.
I had not smiled upon Ethan though rich with
his gift I came,
And his was the highest voice of the bards that
had cried my name.
Tall and agile he was, but little he stood
beside
The bard with the crisp curled locks whose gaze
was open and wide
Out of frank blue eyes that feared not, and
chanted lofty and loud
In their chorus Teffia Teia, and struck his harp
with a proud
Long sweep of the strong white fingers.   His
song ran into my blood,
And its voice is long remembered, as a lonely
tower in a flood.

(2)   My heart hath waited for thee, Teia,
My heart hath waited for thee long.
Though Egypt's sun adore thee, Teia,
My heart is as a hearth more strong.
It shall hold thee, help thee, keep thee,
    Teia,
It shall love thee from this first bright day,
In its radiance fold thee, steep thee, Teia,
When it flashes in the snowstorms far away.
        Green Eriu smiles to meet thee
           Teffia, Teia.
        Her bards are come to greet thee,
           Teffia, Teia.
          With the homage of her love
          That thy crescent smiles above
           In the mirrors of the bay.

My soul is yearning to thee, Teia.
My hands are yearning towards thee now.
Though Tarshish and Pensavel woo thee,
    Teia.
Eriu shall not cloud thy brow.
It shall fold thee, feed thee, fill thee, Teia.

It shall stay thee where the white waves leap,
In thy weeping it shall still thee, Teia,
In thy midnight it shall watch thy couch of
  sleep.
      Its reverence shall be on thee
        Teffia, Teia,
      As a hallowed light upon thee
        Teffia, Teia.
      As the glory of the morn
      Shines upon thy crescent horn
        O'er the emeralds of the deep.

They ceased ere they reached the land, and lo,
    he hem of my vest
Had fallen out of my hand, and the sun that lay
    on my breast
Flashed in their eyes, and they started apart ; but
    the stronger bore
My form in his arms one moment, and set me as
    light on the shore
As I might lay down some blossom, sweet-scented,
    which tenderwise
My lips had touched ere I set it more far to
    gladden mine eyes.

Chapter XIII

(1) Of Tephi's rest at Pen Edair ; (2) She telleth of her person and of her state in going thence unto the Place of Assembly of the men of Eriu ; (3) Tephi rebuketh the priests; (4) Their idol is broken by Ethan.

Now the chiefs of that place and Ath Cliath
    cried my name from their lips,
And a seaman's shouting rolled like thunder
    around the ships
In the speech of the mingled peoples, but
    "Teia" was most their shout
As it was beneath the rock of the Gate. Then
    girded about
By a throng of bright-eyed women, green-tuniced
    and wreathed with green
I was raised aloft on a seat, and carried like
    Egypt's queen
By chieftains in double rank past Edair's piteous
    tomb,
(Edair, Eglaeth's daughter, that died in her
    husband's doom.)
Up the steeps of the Pen to the Cahir of Crim-
    thann, chief of the fights,

Thereafter for and against me in things that the
 Lord requites.

He and his chiefs went before us clearing with
 spears our road,

Their helmets starry with sunset, red suns in the
 locks which flowed

Far down on their crimson garments. Mine
 eyes were dazzled with these,

And I turned and looked behind me, and found
 contentment and ease

Amidst them that followed after, and foremost
 with golden hair

Broad brow and clear bright vision, I saw the
 harper that bear

Me out of my ship, and by him strode Ethan
 agile and dark,

With a flame of fire on his cheek, and fire in the
 eager spark

Of his flashing eyes upon me. Of the bards
 there came fourscore

In green ; then a chosen band of Elatha's men
 from the shore

Came next in their varied raiment, the purples of
 them that sold

The Tyrian wares, and scarlet and azure, whilst
    ruddy gold
Gleamed in their belts and brooches, flashed from
    their helms of brass
Like a marsh-flower mead. Behind them followed
    a mingled mass
Of folk that wore scanty garments waving aloft in
    their hands
Fair wreaths and branches of oak trees, or flut-
    tered on sticks gay strands
Of woollens in tattered ribbons, as bright as a
    barley field
When it whiteneth unto harvest and the husband-
    man guesseth its yield.

(2) Such was my state at my coming. My daughters,
    if ye set store
To hear of a woman's presence, and the garments
    your mother wore
At her welcome ;—little of stature, and slender of
    limb was I,
Being white, not red of my colour, like a stalk of
    nodding rye.

Upon midnight braids of my hair did my argent
    crescent shine.

My throat's thin ivory column poised 'twixt the
    wings divine

About Pharaoh's wide blue heaven; whilst the
    sun of Eileen beneath

Took roses of rosy sunset. On the hems of my
    veil a wreath

Was broidered with gold, and wings of shining
    insects whose name

I knew not, sea-blue below, but lit with an
    emerald flame;

Which veil was long and fragile, as spun out of
    gossamer

By fairy looms of the dawn; and this was the
    gift of Ir

Who had brought it out of Caer Hayo, and said,
    in a furthest land

Of the East, witch-women wrought it in caves
    with a moistened hand,

And withered their eyes in working its whiteness,
    whiter than wool

Or fairest linens of Egypt. Where this veil had
    been folded full

To my form, I fastened and bound it with a
  serpent about my waist
Of fine gold, very precious.   Now in that girdle
  was placed
A sprig of a herb of Eriu, three-headed on every
  stem.
Cendrige, my people call it, and much it is loved
  of them,
As the charm of their fair green island.   This
  those bore forth in their hand
That brought me on cloaks through the ripples,
  and set my feet on the land.
Now this had been placed by the foremost, the
  bard on my right hand side,
But I knew not the charm was with me till I
  found it at eventide
When I couched in the booth by the fortress.
  Next morrow at early dawn
When my women arrayed me for journey, I saw
  it, and scorning to scorn
A bard that had given such welcome, set it again
  to the clasp
Of my serpent ere Bres came thither, and lifting
  me light in his grasp

G

Placed me high on a jennet, snowy, wild-eyed and
    still,

But therewith tall and stately, and so we paced
    down the hill

And out through the fair green grasses, with Bres
    still near at my side,

And his cohort of captains by him wherewith he
    was wont to ride,

And the bards behind us on ponies that sat with
    their harps to play

And move us with mirth and music what while
    we went on our way.

Now Ethan was ever foremost, and sweetest of
    all was his song ;

But I looked in vain for his fellow, with purpose
    that held no wrong

Of repaying his charm with a golden ring, but I
    found him not,

Marvelling wherefore he tarried ; yet my cendrige
    was not forgot

When we came by an easy journey next morn unto
    Crofinn's croft,

Where at the land was assembled, for there the
    grasses were soft,

And many horses might pasture, with cattle and
    flocks for meat.

Here the chiefs of Eriu had portioned themselves
    their seat

On the banks round the croft of Crofinn, and there
    each set him a booth,

And they met on its central greensward where the
    level was clean and smooth

For choice and converse amongst them upon
    Eriu's hap and its weal,

In a three weeks' truce wherein the tongue was
    lord of the steel

Throughout all coasts of the island.   Now this
    truce was for two days yet,

When one short hour after dawn, through meadows
    that still lay wet

With the dews I came to the croft as a queen
    with my following,

For unto that day the island had never bowed to
    one king,

Though high chiefs ruled in Usna and Caseal and
    fair Emain,

And in many duns and cathirs fortressed in forest
    or plain

Or on hilltops.   Each tall landmark crowned with
    their strongholds stood,
And the lords did that within them that seemed
    in their own eyes good.
Now the cry of the land was bitter, for most of
    the chiefs wrought ill
On their landsmen as on their foemen, and each
    by his strong-walled hill
Held cattle plundered of either, until the forces
    which cling
In clanship were severed amongst them, and the
    aires cried for a king
To hush their feuds and to pluck the husbandmen
    from the mire,
And the bards of the land were with them to
    yield them their heart's desire ;
But the priests of the gods against them.   Yet
    some of the priests that knew
The God of the Hebrews helped them ; but these
    were a chosen few,
And the priests of the heathen many, well skilled
    in the ancient lore
Of Criden and Baal and Samen, and many an
    idol more

Whom their fathers knew in Canaan, and the
    June morn filled with heat
When I heard their trumpets blow as the priests
    came forward to greet
Her that was hid in the Temple; yea, in its
    inmost shrine
Was held with the graven tablets, and the scrolls
    of the Law Divine.
These that came in white garments.   These with
    a frenzied tread
That whirled upon desiul* circles !   Had not
    my spirit bled
Before such in the house of David ?   How might I
    greet them here ?
I was weak, the might of the Lord prevailèd over
    my fear,
And I rode in His wrath against them.

(3)   " Ho, ye that have eyes to see,
    Ho, ye that have ears to hear with, keep silence
        at sight of me,
    And my voice from the Lord," I cried, " for Baal
        is broken of Bel,

* Sunwise.

The twain shall be broken together.   They sink
    to the nethermost hell.

A flame hath descended on Zion.   God sweepeth
    with wings of fire

The House of His habitation.   He sendeth hail
    upon Tyre.

Zidon and Gath are broken, Ephraim led away,

Samaria lieth fallen, and is as an heap this day

Because men whorèd with idols.   Shall idols
    come forth to greet

Her that the Lord hath kept, that dwelt by His
    Mercy Seat.

Your dances and fires He hateth.   Behold, the
    face of the Lord

Is a sun that shineth in darkness, His tongue is
    a flaming sword,

Let Criden and Baal be broken, devourers, and
    blind of sight

And empty of help for all that sink in the womb
    of night.

Yet the great or little prevails not when God
    ariseth in wrath,

With a pebble-stone from the brook he layeth the
    might of Gath."

(2) E'en at my word a pebble sang by mine ear and
    smote

Through the open mouth of Criden, and broke
    his head from his throat,

And his breast was shattered also.   Swift on my
    own swift speech

Was Ethan's deed upon Criden, for all that the
    prophets teach

Was known of Ethan, our Hebrew speech, and
    our father's deed.

He smote as my father David.   The Lord had
    answered my need.

Now the image he smote was hollow, and held in
    a secret hold

The gifts of the blind and foolish, their rings and
    the stars of gold

Which the priests said went to his dwelling, but
    now his falling revealed

From the hiding place of his belly, and scattered
    o'er all the field,

And all were amazed and angered; and men
    called out upon Sri

The son of Ezra their wisest, to interpret my
    word, and why

Their idol was shattered before it, for silent
   amongst the band

Stood Ethan, and none beheld when the stone
   flew forth from his hand,

Their eyes being set upon me; and wherefore
   that image fell

When my wrath was laid upon it not they that
   bear it might tell.

Then Sri the son of Ezru, a lover of better
   things,

Set forth my speech in their tongue, and the
   strifes of our former kings,

How Saul the mighty had fallen when idols led
   him astray,

And how from the house of David God's curse
   was taken away

For a space, but was sealed thereafter. Now the
   priests were angry that heard,

But the common people listened, and many
   hearked to his word,

And some of the chiefs and the most of the
   bards. Amongst them a cry

" Daouda, Daouda hath smote him," arose at the
   words of Sri,

Telling how David had smitten whilst yet a
 youth with the flock

The giant, mighty in war, with a stone of the
 brook, a rock

The cornerstone of his house : and the shouting
 " Daouda " grew

When he told how the Lord of Hosts descended
 in flame anew

On the Seat that he brought from Kirjath to set
 in Jerusalem,

The Ark, the Holy of Holies, which went with
 the tents of Shem

When Israel came out of Egypt.   Sore were the
 priests of Baal,

But the people cried out against them, and
 praised me that heard this tale,

So their wrath kept silence before me, and
 turning they went again

Till we passed the banks of Crofinn, and entered
 the little plain

Wherein the chieftains assembled.   An hundred
 princes and eight

Of Eriu waited my coming ; each with his proper
 state,

His druid and bard and champion ; and all stood
    there on their feet
Save I, who with Bres at my bridle, rode forth on
    the sward to greet
The lords of the high assembly, who hailed me,
    child of their isle,
And queen of the house of their fathers, and so
    without thought of guile
I unveiled my face before them, and spake to
    them gentlewise
My thanks for their greeting and favour, but that
    which shone in the eyes
Of many chilled me before them; so, icy in
    pride, I rode
Before Sri, and Bres and Ethan, to enter the fair
    abode
Which these had built for my coming, whitewood,
    well carven in scrolls
Of serpents, whose hinder part in an endless
    ribbon unrolls.
Its door was a woollen curtain of green with a
    scarlet hem,
And Sri on its lintel fastened the name of
    Jerusalem

Writ in Hebrew in brazen letters, and set on its
      posts a sign

That none but the maids might enter the booth
      which was named as mine.

Therein I rested at noonday, and ate in the
      failing light,

But had little sleep thereafter, and watched the
      most of the night :

For the looks of the priests misliked me, and the
      hungry eyes of the men

Of Eriu searched upon me, as eyes of wolves in
      their den,

Till my heart was water within me, troubled and
      sore afraid.

Then long in the long night watches to the Lord
      of Zion I prayed

To deliver my soul from evil, my limbs and breast
      from the grip

Of a wolf, and the High One heard me, and
      caused not my foot to slip.

CHAPTER XIV

(1) *Tephi telleth of the departure of the Man of God;* (2) *She is brought on the morrow into the assembly;* (3) *The lot of Baal falleth beside Balor of the Mighty Blows, and upon Bres the son of Elatha.*

(1) YET my troubles that night endured, and I
    longed for the Prophet's aid,

For I loved him e'en as I feared him, as an
    infant standeth afraid

Of a father strong and silent, yet knoweth his
    help shall come

From thence if the wild beasts fright him, or
    robbers seek to his home.

My sons, ye enquire of the Prophet. This sure
    word I bid ye to know,

Mark well the way of the chosen, but seek not
    whither they go.

Pause on their word and ponder though at times
    ye may not mark

Their message. The eyes of the holy behold a
    light in the dark

Of Tohu and Hinnom wherein their path hath
    been set to go

Through night.   On their heads are ashes.
   Their garments are rent in woe.

Lamentation is with them and terror, till the
   terror be overpast,

For they grope after God in Tohu till they find
   Him and hold Him fast.

I dwell not now on the thing which shall in this
   book be told,

How hereafter dimly mine eyes should the Friend
   of the Lord behold.

He sought not pleasure of greetings, or tables
   of wine and meat,

Or to listen to mirth or music, or to sit in the
   highest seat,

Or behold me in marriage garments : but set his
   feet in the way

Of the Lord where'er he led him.   This only
   therefore I say,

That when we had left Pensauel, drawn nigh to
   the land of Gwent,

He parted his ships from amongst us, and none
   knew whither he went

With the sacred things of the Temple, and none
   may utter their tale,

For his sailors were men of Ham the last whom
the Temple veil

Shall leave in the dark ; and these that sailed on
the western track

With the Prophet passed into night, and ne'er
out of night came back.

Of the sacred things I know not. The Lord
stays not to discern

The place of His habitation, whereunto my sons
shall yearn

In the days that dawn hereafter ; but lo, ye
have seen the Stone,

The Stone of the Corner remaineth. It shall not
be left alone,

When Jacob knoweth his birthright therein shall
his boasting lie,

And in many lands and islands my seed shall
have praise thereby.

There was one beside the Prophet mine eyes
were fain to have seen.

The morn that I came to Crofinn, I watched for
the cloak of green,

And the strong straight bard that wore it, as one
looks for a trusted friend

Amongst strangers.   Perchance he guessed not.
Perchance he might not attend.

(2) On the morrow came Bres with Sri to lead me
forth to the ring
Wherein were the chiefs assembled to hear men
cry for a king,
But each man envied his fellow, and each with
an angered mood
Had answered the bards and aires that spake for
the common good.
My place was set me amongst them, a seat upon
Jacob's Stone
Drawn thither by two white heifers, and draped
around as a throne
With a golden cloth of Zidon.   Now, as I was
set thereon
A cloud drew back in the sky and upon me the
bright sun shone,
So folk marvelled of me and this sunshine, and
thus it was foolish talk
That I held the sun at my bidding, setting paths
for the clouds to walk

At my will, and I own I had joy, for I cast on
    the Lord my prayer

In the night, and now in the day he had lightened
    my load of care.

Now this same day was an high day, the topmost
    peak of the year

Is the night that follows after, when angels and
    souls appear

Unto many, yet here the druids had mingled its
    boons with harms,

And setting their hearts on women delude them
    with evil charms.

(3) A feast being set to Baal, his priests drew nigh
    ere the noon

With a message brought from his altar that the
    king be appointed soon

As this one day was propitious. The bow of
    their god they brought,

That by this an arrow sent sunwards should
    name the king of his thought,

So we all drew off a little to the banks and stood
    to see

How the highpriest bound his eyes, and drew the
    bow from his knee

Where he lay supine, and the shaft sped upwards
    to seek the sun,

But an East wind struck upon it ere the height of
    its flight was won

And bore it beyond the circle where it fell full
    nigh to the ranks

Of Balor, lord of the Islands, where he watched
    with his men on the banks,

And his firbolgs shouted for Balor, but the priests
    were troubled thereby,

For their spells were within the circle ; so another
    quest of the sky

Was made, and it touched the circle, where
    nearly it struck down Bres

In whom was a hope of Eriu that ever grew less
.   and less,

For when Nuadh was maimed in battle, men held
    that his strength was stayed

From rule of the miledh of Dan, and a pact unto
    Bres was made*

* For seven years.

H

That he should be named chief captain, if so he
    would save the land

From fomorcs coming by sea, and chiefs of the
    scattered band

Of firbolgs in Man and Arran, so this for that
    time was done.

But he gathered Eriu's tribute, yet gave its gifts
    unto none,

Neither called he feastings or music.   His heart
    was empty and bare,

Though the strength of his limbs, and his beauty
    of face, and his golden hair

Snared foolish matrons and maidens.   Yea, deep
    in his heart was guile,

And women loved and men hated his presence
    throughout the isle.

Now the arrow struck through his cloak, and
    pinned it unto the ring

A handbreadth from Nuadh's high seat, and
    many acclaimed him king,

That was chief of the miledh of Eriu ; but the
    priests had marvel thereat

If the shaft were within the circle.   Moreover,
    the place where he sat

Was apart, and the shafts of Baal were counted
    not to the man

But rather the beth of his ensign. Moreover,
    they loved not Dan,

Of whom was his mother, and whom he spake for
    in Nuadh's room ;

Whose hand was severed by Sreng the son of
    Sennchan, whose doom

Thereafter the scribes have written. Now Dian-
    cecht, wisest in art,

Had moulded a hand in clay wherefrom might be
    hurled a dart ;

And Creidna, the cunning smith, in silver
    fashioned the same,

So now the hand of Nuadh flashed with a starry
    flame

As he rode amongst his miledh, and many that
    loved him well

Sware that the seat of Nuadh was grazed when
    the arrow fell.

Chapter XV

(1) *Sri, the son of Ezru, calleth for the bow of Sampson, which*
is given to Ethdan,* son of Bœthlam, who shooteth the
first arrow unto the Stone of Jacob ; (2) *A second arrow*
is shot, and findeth the seat of Eochaid Garbh Mac
Duach ; (3) *The sun betwixt the horns of Baal is smitten*
by the third arrow, and Sri, the son of Ezru, maketh a
psalm thereon.

(1) Now the priests and chiefs of the land debated a
    threefold choice

And a doubtful, striving greatly, till Sri with a
    mighty voice

Cried, till they heard. "Not yet is the curse of
    this kingdom stayed.

The sins we have sinned to Baal shall yet at our
    gates be laid.

His arrow hath pointed Bres, it hath fastened his
    garment's hem,

In the folds of his cloak shall Canaan set fires in
    the booths of Shem."

These things cried Sri the silent ere shaping his
    theme anew

* Breasal Ethadan Mac Eochaid Bœthlaim—too long for verse.

He said "the arrows of Baal seek sunset or fall
    askew.

Seek we shafts that are truer. Is there not in
    our midst the bow

Of strength, the shafts of the mighty? Where
    Dan goeth to and fro

The bow of his judge is with him, It dwelleth
    amongst us here.

The merchants of Gath and Japho draw back at
    its name of fear.

Have we never a champion of Dan who may
    string its strength to his will?

Is the spirit of Sampson weak to speed the shafts
    of it still?"

Then Ethdan the son of Bœthlam thrust through
    the ranks of Dan,

Of all the sons of the island this was the broadest
    man

Of shoulder and girth of limb, if somewhat slow
    of his feet.

He called for the bow of the mighty, and strain-
    ing back from my seat

He bent it. Mighty the string wherewith that
    bow must be strung,

A finger of sinew to armbreadth of yew, but at
    last it sprung
To the cleft with a stroke like an axe when it
    striketh an oaken beam,
Whilst the flesh upon Ethdan's arms sank like
    waves on a stream.
Then swift to the circle's centre he sped him and
    laid him down,
Setting his feet to the yew-mast. In a moment
    the shaft had flown
Straight into air till we lost it, and then in a little
    space
Straight out of heaven it descended like a beam
    of the sun on the place
That was mine, the Stone of Israel, yet hurt not
    the Stone at all
For the head's soft gold spread forth a sun at the
    arrow's fall
On the greywhite pillar of Jacob ; and joy upon
    all men came
When they saw the altar of Bethel alight with
    that golden flame :
And the priests of the gods bowed down, and
    covered each man his face ;

And the chiefs of Eriu moved in wonder before
    that place ;
And little they spake, but set me thereon ; and
    lo, I had grace to speak
In their tongue, and my heart was great, though
    my voice was little and weak.
"Ye Chiefs of this island, hear me. The might
    of the Lord is known
In shadow, but light is rising, and grace to a
    handmaid shown
Who watched and prayed in the darkness. He
    leadeth her by His ford
To sit in a fair green pasture, with sheepfolds
    and oxen stored.
A shepherd was David my father. God gave
    him a charge to keep
Which he brake not, to feed His cattle and sever
    the goats from the sheep.
Me, that am David's daughter, he maketh a shep-
    herdess
Who amongst the sheep of Eriu shall know none
    greater or less.
The sun that descended hither shall be as a light
    divine

Whereby to search in your pastures, and know
    my sheep from the swine,

For the unclean beast is with you." Then Sri
    that stood at my side,

Passed up the banks and turning, to all the people
    he cried,

"The Queen of the East hath spoken. Is there
    one her word to gainsay?

Let him dwell with the swine, for God hath sent
    us a Queen this day."

Then Tuirbhi the smith sprang forward to catch
    at Sri by the arm,

But Sri smote straight upon him and wrought
    him a deed of harm,

For he fell by the banks on his ancle, and his
    craftsmen bore him away,

And his leeches bound him badly, and lame he
    went from that day.

(2) Then Ethdan the son of Bœthlam, cried "there
    were arrows three

With the bow of strength, and the first hath
    sped; but I ask of ye

That be wise, shall I speed these others? The
    one hath a silver head,
But the other is somewhat crooked and beareth
    a bolt of lead."
Then the priests drew nigh giving counsel, and
    the most spake well of the thing,
So we left the plain as aforetime, and forth from
    the mighty string
The second shaft flew upwards until it was scarce
    discerned.
Like a star it glanced on the cloud, and then
    unto earth returned,
Smiting an oaken settle which no man had used
    that morn
But sideways lay on the ground, and grazed it,
    and cleft a horn
Of silver therein, and smote into earth, and a
    question rose
Of that seat but no man claimed it, its chieftain
    was not of those
That sat in that day's assembly, and pain sank
    into my heart
At that long carved cleft of silver, which stabbed
    with a sudden smart.

(3) Now Ethdan fitted again the crooked shaft to the
  bow

Which sped on a snake's path outwards, like a
  hawk when it striketh low

But swiftly above the gazers, till the pillar of
  Baal it found

And struck the gilded sun 'twixt the idol's horns
  to the ground,

Bearing it into mire in the place of the swine
  behind,

Wherein they lie to this day. If ye search, ye
  shall surely find.

Now when they beheld this token many priests of
  the idol fled.

Through revilings amidst the people, and tore
  their wreaths from their head,

Gashing the flesh of their bosoms, and hid themselves ; but a few

Remained in the ring with Ethdan. Then Sri
  that was wise and true,

Though his knee had bended to Baal, cried out
  on the Lord for aid,

Forgiveness, counsel and blessing, and a psalm
  of repentance he made

Which the bards took up in chorus, singing it
    hither and fro

From the priests to the kneeling harpers, who
    sung to a music low.

"We walked in clouds of the night.—Our eyes
    are opened by Thee.

We look unto heaven and see.—Yea, we awaken
    to light.

Thou knowest our blindness, oh God.—Let thy
    forgiveness prevail.

Sorely our sin we bewail.—Let not thy spirit
    record.

We are troubled of heart in thy presence, oh God.
    —Yea troubled sore.

Thine angels vex us, thy saints abhor.—We are
    struck with Thy rod.

Thou sendest us consolation.—Therefore Most
    High we give praise.

Thou hast chosen a day of the days.—Thou
    sendest a queen to this nation.

Thou, Lord, art a righteous King.—Out of heaven
    thou givest favour.

Let our song be of sweet savour.—Lord, in Thy
    praise we sing.

Chapter XVI

(1) *The seat whereon the silver arrow had struck is known for
the seat of Eochaid ; (2) Tephi resteth thereon when he
cometh, and giveth her love unto him ; (3) the Queen
taketh Eochaid as her husband ; (4) the spirit of prophecy cometh upon Sri the son of Ezru.*

(1) Now even whilst they sung a cry rose round about
  The shrine of Baal, the commons made a mighty
   shout,

  Hauling at ropes and girdles till the lofty pillar
   crushed

  The turf, and for a breathspace the sound thereafter hushed,

  But Baal avenged not aught, men seeking each a
   stone

  Wherewith to bury Baal, whose resting place is
   known

  Beside my house at Teamur.   Then Sri and
   many more

  Gazed nearly on the furrow which the second
   arrow tore

  In the oaken seat, and Ethan who departed for a
   space

Drew thither, and one asked him was not this his
  chieftain's place,

And on that question Ethan raised to mine a
  face of flame

Till my brow was veiled before him finding
  searching prayer and shame

In the gaze he set upon me ere he answered to
  them "Ay,

This is Jochad's seat and hitherto my songs were
  heard thereby."

Then Sri questioned further wherefore did the
  Heremon * eschew

To be with them on this high day, and the brow
  of Ethan grew

Pale and red as he gave answer, "'Tis the third
  day since some cause

Which I know not drew him homewards from
  Pen Edair." At his pause

Fell my veil, and full upon him was my gaze, and
  well I knew

That if truth he spoke, it shamed him in some
  thought not wholly true.

* Chief of the landsmen.

Though I spake not, he gave answer in a sudden
      word and swift,
" Read his secret. Thou dost know it." Then
      my veil I did uplift
Once again, for blood ran tingling over breast
      and cheek and brow,
And a spirit quickened in me which I had not
      known ere now,
Some strange gladness half an anguish shook my
      bosom till I swayed
Like to fall, but Sri upheld me and he set me in
      the shade
Of the arched highseat of Jochad whereupon the
      arrow fell.

(2) There I rested till a voice out of the distance
      seemed to swell
Drawing nearer. " Jochad, Jochad," but as in a
      trance I lay,
And mine eyes were blind and misty, till a sudden
      golden ray
Fell upon them with a sparkle and a light to
      overwhelm

Every mist. Grey eyes and fearless gazed beneath a golden helm!
So my soul's sun dawned upon me, and I rose
up from my seat,
Whilst the sun bowed down beneath me plucked
a cendrige by my feet.
White I stood as stands a statue when he touched
the new plucked leaf
To the withered at my girdle, kneeling still, but
still the chief
Of my stature, and the crescent which upon my
brow had rest
Was beneath the leaf he gathered when he set it
in his crest.
Stark he knelt in homage pleading to my crescent
where I stood
Icy cold, till some strange Summer thawed away
my Winter mood.
Weak I grew and blind and dizzy in that new-
born Summer drouth,
And my hands stayed on his shoulders, and my
lips just passed his mouth,
And a cry was all about us in the dancing shapes
around

Moon and sun are met together, and this place is
  holy ground.

(3) My bridegroom, my chosen, my strong one, in
    whom my soul had delight,
  My feet were by thine, my hand was in thine, as
    they led us to plight
  Our faith by the Stone.   My heart was thy heart,
    My will was thy will,
  When Sri and the priests spake with us, and
    bade our souls to fulfil
  The vow of the lips by vow of the soul and swear
    with the Soul
  In sight of the people and priests and scribes
    that stood to record
  Our oath of faith with people and priests and
    chiefs as a pair
  That God made first in the land, to have it in
    heedful care
  And seek not ourselves but Eriu.   The words of
    that sacred oath
  Were mine, but I know the Spirit of God had
    fallen on both

For his day of days, being joyous thereat in a
    waking dream

Wherein all faces and garments danced in one
    sunny stream

Of eddying light, one only resting stalwart and
    tall,

For though many great chiefs were round us he
    stood the first of them all.

After that oath I stood calmer, and watched with
    a careful eye,

When the oldest priest of Eriu set in the hands of Sri

A vessel of alabaster that once in the Promised
    Land *

Was shapen and graved with the names of God by
    its maker's hand.

Its oil had been pressed from the harvest of the
    garden o'er Kedron's brook

Whereon mine eyes in childhood from my window
    were wont to look,

Being perfumed with nard and cassia, most
    precious. Then Sri drew near

To anoint me, but I stood up on my Stone, and
    said without fear :

* Tir Tairngre.

I

On this stone I am set for ever.   In Egypt
    anointed queen
Of the Hebrews.   My throne in Jesse hath come
    to these hills so green
For a little space, ere it wander, but wheresoever
    it roam
Jesse shall seek and find it until he come to his
    home
In the City of David wherein his sons shall rule
    upon earth,
When the house of the Lord be builded with
    praise and blessing and mirth.

(3.) Then Sri, being moved, forbade that my hus-
    band's seat be with mine,
And prophesied of us saying: "This shall be
    kept for thy line
And for thee ; but he that is by thee standeth on
    Eriu's sward.
It is his by birth, and hereafter, this island shall
    name him lord
Of its people to be their leader, and shape their
    counsel in war :

But thou art of Isaac's children the guide and the
　　crescent star,
Wherein thy children shall shine, till the full
　　round circle shall beam
Of that orb wherewith the moon at her first
　　appearing doth teem.
He that is chosen amongst us, He shall be great
　　in thee,
And thy sons that shall be after.　Is not his lot
　　to be
A father of thrones and kingdoms?　This is the
　　name he shall bear.
In the tongue of this people his title is Eochaidh
　　Ollothair,
Eocaid, Sire of the Great Ones ; these sons of the
　　land which is great
Magh Mor, or of Og, the holy, that they learn of
　　their own estate,
And yearn to the promise, and David bless them if
　　this they know
That holiness unto the Lord is their greatness
　　wherein to grow."
Thus then spake Sri, whose silence to God was
　　on all men's tongue,

For the mouths of them that knew him, since in
Ezru's house he was young.

Ezru that fled out of Ghor, * when Asshur came
with his bands,

And ere he came unto Emain taught wisdom in
many lands :

But the mouth of his son was shut till his spirit,
nurtured of prayer,

Spake with the Spirit of God which worketh in
stones and air,

And whispers by reedy waters, and moves in the
mountain's shade,

And knoweth the inward parts, and wherefore
man's soul is afraid.

Now men marvelled much upon Sri, having
feared him and called him wise

And wary, but said that he feared neither spirit
nor prophesies,

Having taught as the scribes from rods, and the
teachers from ancient rules,

Being learned in many tongues, and chief of the
poet's schools,

* Fr. Gorius.

Fearless but scant of speech, and though wisdom
dwelt with his word,

To this day his voice was silent when men spake
the praise of the Lord.

I beheld the people's wonder, and looked upon
Sri and knew

The mantle I oft had seen, and his word as a
prophet true.

And was glad in the Lord as my helper, whose
word should be held of me

As his who had led me from Egypt and helped
my paths in the sea.

CHAPTER XVII

(1) *Maistiv,** *the sister of Eocaid greeteth Tephi, and telleth of
her brother;* (2) *Eocaid speaketh of Ethan and Bres ;* (3)
*Ethan, the son of Becelmus maketh a song, whereat the
heart of Tephi is softened towards him, beholding much
good in the man.*

(1) Now soon my heart contracted, for a damsel
stately and fair,

Broad-browed, full-eyed, and gracious beneath
the crown of her hair,

---

* The exact relationship of Maistiv to Eocaid is somewhat
doubtful ; she may have been his aunt.

Large-limbed and nobly shapen, tall to a chieftain's
    height,
Drew from the throngs before us, and now with
    a queenly right
Took my bridegroom's head in her palms and
    kissed him upon the lips,
Whilst cold went through me which passed from
    heart unto finger tips ;
But my husband smiled, and said, " My queen,
    yet thy servant's bride,
Behold the chief of thine handmaids, my sister
    Maistiv, whose pride
Is Dan, Achaia and Eriu, who in her give fealty
    to thee
Of the silver stem of Jesse, the golden flower of
    his tree."
Thus shamed I my doubt with blushes, and we
    kissed, and were ever knit
Though golden and dark, as sisters, unlike, yet
    never a whit
Sundered in our unlikeness ; and Maistiv knelt
    at my side
And told me that which gladdened my summer
    of heart at that tide.

But three days since as she wandered with one of
    her maidens near

In the bowers of the woods by Mulach, thinking
    to have no fear

Through the sacred days of assembly, lo, Bennan
    the son of Kain

A foster servant of Balor's with seven men of his
    train

Drew round her and led her with them ; but her
    maid that was nigh had seen

From the hazel brake their doing, and slipped
    from the leafy screen

To ride in haste to Pen Adair.   Then, straightly
    upon her word

Had Jochad taken his breastplate and girt him-
    self with his sword

And leapt to his horse's saddle with three that he
    had thereby,

Who galloped the trail she told of all day till the
    midnight sky

Was sprinkled with stars, and came to the spot
    where Bennan stayed

His course with the setting sun, and three of his
    train were laid

Before them upon their onset, and one as he fled
    away
Was stung by an arrow, but Jochad sought
    not further to slay.
Setting her safe on his horse, which weary, carried
    them back
Unto Mulach, her house, but scanty of patience
    was he till their track
Was westwards in haste to Crofinn, whereat
    much wonder had been,
But now she wondered a sister had drawn him
    away from his queen.

(2) "Ay, sister," said Jochad, "a wonder, and much
    had I longed to remain
If I had not brother or friend, but much I dwelt
    on these twain,
Ethan and Bres my brother. In these I might
    cast out fear
Lest the queen lacked fitting service, or my watch
    of her light be near."
Then he turned upon Bres and Ethan and held
    out a hand to each,

And the first grasped forth at the hand, but
    Ethan slipped 'neath its reach
And knelt till it touched his head ere he kissed
    it with downcast face.
Then smiled my husband in chiding, and raised
    him up in his place
And kindly questioned his gaze, and said, " Is it
    well that thou
The chiefest bard of Eriu to a yeoman of Eriu
    bow ?
Thou castest down and thou raisest up. Our
    glory in death
Is left to the bards that fill our ghosts with un-
    dying breath
To rehearse our deeds to our children. Oh poet,
    make us a lay
As glad as this hour is joyous, upraised as our
    hearts this day."
Then Ethan said, " My lord and my king, my
    spirit was dead and mute.
I was cast in the mire till thy coming. I have
    broken the strings of my lute.
I have sinned and done great evil, and how may
    thy servant sing ? "

And my bridegroom frowned, but I took from
    my finger my golden ring
Fired with a heart of ruby, and said, " If a poet
    know
His evil, he eateth knowledge, and knoweth of
    good also.
I give thee a bane of serpents.   Take this as a
    charm to part
Thy soul from venom, such magic is stored in
    my ruby heart."
He set my gift to his lips, and never a harp he took
But music out of their parting poured like a
    running brook
As he sang the bridesong of Crofinn, glad as that
    hour was glad
Are its words, and its fame is with him, but at
    whiles his eyes drooped sad
On earth ; then, lifting again, they brightened
    clear at my sight,
And turned on my bridegroom also, and were
    honest and filled with light.

(3)   What shall I sing thee,
           My mistress, my queen ?

What may I bring thee?
Heart's blood I would wring thee
  Were this not too mean.
Thou hast bid me to sing
  My master, my lord.
From thy servant, oh, king,
Take this, the queen's ring,
  It is all of my hoard.
This ring had its heart
  Of the Lord, the Most High.
By its magic of art
It shall throne thee apart
  In the midst of the sky.
Thy place under heaven
  Is near by her seat,
From dawn unto even
Thy foeman forgiven
  Shall kneel at thy feet.
The Lord, the Bestower,
  Gives gladness to thee.
Betwixt higher and lower,
  He builds thee, His tower,
  For this isle of the sea,
Whose lowly shall love thee,

Whose lofty bow down,
Whose priesthood approve thee,
Yet this gem set above thee
  Shall be thy renown.
To thine honour give heed
  And thy manhood with man,
Being noble in deed
Being chosen in seed
  Being princely with Dan.
Yet the light of thine eye
  Thy knowledge, thy truth,
Are faint in the sky
When thy moon rideth high
  O'er the bosom of youth.
The magic she maketh
  Is silvern and pure.
From the heart that she breaketh
A spirit awaketh
  With strength to endure.
Receive this, my king,
  With sweet spirits well stored.
The queen's heart, her ring,
Save the lays that I sing
  It is all of my hoard.

(4) We heard, and Jochad rejoicing, gave him his
      finger ring

Golden, with fair bright pearls such as men of
      the Sgiath bring

To our north coast; yea, and I gave him no
      jewel or golden gem

But the olive twig my fingers had plucked
      by Jerusalem

To keep my heart in remembrance. So fled
      the cloudlet away

That in all the light of Summer had shadowed
      my joy that day.

Then the priests went desiul * round us thrice,
      and chanted a charm

To stay our steps by each other, and fence us
      from outer harm,

But I know that we needed naught in our circle
      of hearts complete.

So went we in to the feast, where I sat in the
      highest seat

Betwixt my husband and Maistai; and Ethan
      sang to the guests,

And Sri gave blessing upon us before we went
      to our rests.

          * Sunwise.

Chapter XVIII

(1) *Balor the descendant of Neith goeth homewards angry ;* (2)
Tephi sitteth in judgment ; (3) *Cairbre the son of Etain
maketh a song against Bres the son of Elatha.*

(1) At the dawn we heard how Balor of the western
islands had fled

By the slope of the chariots homeward. I had
heard his horse's tread

And his wheels of iron ere dawn, and marvelled
of what might move

With that sound and quaked in the dark, but
the bridegroom spake words of love

Which builded my heart in strength, and spake
of those things that I

Might work in this land of the ocean, if the God
of my sires was nigh

Unto me as to Moses in Egypt. And thus in
this far off strand

My heart might be cheered within me with sight
of the Promised Land.*

He had heard the songs of Zion, and the common
folk in prayer,

---

* Tir Tairngre.

Named its name as a charm, and knelt with their
    faces there,

Not sunwise as the priests did ; and his spirit was
    sorely grieved

When I told him of Zion's fall, and greatly his
    heart believed

In the Lord, and he prayed that idols might
    forth from our land be cast,

And joy return to Moriah, and its sorrow be
    overpast.

(2) When we went from our booth at the morn, I
    was led to a little hill *

By the banks, whereon was my seat ; that before
    the people, my will

Might be seen and known of many, and Eriu
    learn my word.

Which Sri, son of Eschmun the scribe was set
    by me to record,

With Aci, son of Alghuba, as herald to shout my
    choice,

Or proclaim my goings before me ; for his was a
    mighty voice.

* The royal hill of the judgments at Tara.

In warfare or peace, save Ethan, was no man
        broader than he,
And these twain I set together for truth and
        service to me,
With En, and with Sri, and with Ogma, my
        husband's champion and friend,
My almost brother, for these were faithful unto
        the end,
And helpful in my beginning; also Nuadh, the
        brave old man,
Who all the days of his youth was chief of the
        host of Dan,
And led the miledh of Eriu, ere his hand was
        smitten in fight;
Being first to kneel at my feet; and that old
        man's eyes were bright
And his strength not yet abated. He spake as a
        man of war,
That his knees were stiffened with age before
        men, but queens led far
And their followers never wearied; so, smiling, I
        give him thanks
For himself and his band of Dannites, and a
        cheer went up from their ranks.

Many a chief came after, and Crimthann came
    with the rest,

And Bres, and my husband also. It irked me
    much that his quest

Was to sit in my sight before me, yet ill example
    had been,

If one alone unquestioned might break the state
    of the Queen

Being set in judgment on all men. Full soon
    my judgments began,

For a chieftain of Crimthann's came with claims
    on a husbandman

Whose few sheep wandered astray, and ate three
    days of his land

Ere the aire found them. Then Crimthann
    standing forth from his band

Claimed the sheep for the grass ; but I said "the
    flock and the field

Have titles, but know ye not that each hath its
    proper yield,

Take ye three fleeces then, but leave the aire his
    sheep." *

---

* This judgment belongs of right to Cormac Mac Art.

K

Then e'en Crimthann laughed aloud, and sware
　　that my laws were deep,

And fleeces should go for the grass.　So Aci
　　shouted aloud

This judgment, and praise and laughter arose in
　　the mingled crowd.

(3)　Then a weighty matter beset me whereat I was
　　ill at ease,

Baring my thought unto God, yea, even as on
　　my keees.

A bard of the land stood forward, and bidding
　　the chiefs regard

His song, he chanted "the rights and due re-
　　wards of a bard,"

And rehearsed "the rights and duties and proper
　　state of a chief,"

And then "the customs of Eriu in all that re-
　　gards a thief.

And the shames that await a niggard."　Lastly he
　　spake the grief

Of Eriu in yielding tributes to save her shores
　　from her foes

Without, and within her taxings, and her burden
    of heavy woes

From the chief's fierce guards and firbolgs.   " Our
    miledh " he sang " we keep

As sheep-dogs to guard our pasture, neither
    sheep to feed with the sheep,

Nor mongrels with cheftain's mongrels who snap
    at the lambs in fold.

But these watch-dogs bark in the sun, or snap
    upon flies, grown old,

But Bres, their leader is watchful, he setteth his
    ships by the beach.

His jaws are ever open, he sucketh the tax like a
    leech.

He storeth gold in his chamber, even in every
    house

Of Bres is a treasure chamber, but therein never
    a mouse,

For the tables of Bres are empty.   I passed by
    a house of Bres

Who sat in a broidered garment, and toyed in
    his wantonness

Amongst the locks of his damsels.   His arms
    were laden with rings

Of Eriu's gold.   Then sang I his wealth, and
the mighty things

That he wrought in fight with the Firbolgs ; after
Edlai and Turild were slain ;

And Nuadh wounded of Sreng might hardly the
fight maintain,

How he slew Mac Erc, and drove the Firbolgs,
and compassed about

Strong Sreng, till he gave him pledges.   This
land hath never in doubt

The strength or beauty of Bres.   By land and by
sea we know

Men fear him and women love him.   Why then
is his glory low ?

Save unto foolish maidens the welcome of Bres
is cold.

Save for his own attiring the garments of Bres
are old.

Save on his shipmen's armour he spendeth little
of gold.

At his door is a couch of purple.   His guest is
set on the sward,

At his door the blind and the lame unto prayer
  find scant reward.
On his door are bars of iron wherewith he
  guardeth his hoard.

In his house is neither music nor laughter nor
  sound of feast.
In his house a fierce hound snarleth but never
  another beast.
In his house is neither aire, nor chieftain, nor
  scribe, nor priest.

On his hearth is one small fire, it roasteth a little
  food.
By his hearth a stout wench turns it, and the
  smell of the meat is good.
By his hearth one trencher is warm though he
  burneth but little wood.

In his cave are rusty cauldrons that his mother
  once filled with ale.
In his cave are rotting meadvats, for his bees and
  his honey fail.
In his cave is a broken pitcher, and the whey in
  that pitcher stale.

In his closet are wines of Chittim which even as
rubies shine.

In his closet wine of Tarshish like molten gold of
the mine.

In his closet are precious vessels, and one was
brimming with wine.

For the bard a fragment of bone !　For the bard
the pitcher of whey !

For the bard a seat on a stone !　For the bard a
hovel of clay !

From the bard sour whey, picked bone, cold
stone, for a prince this day !! *

---

* The above, though not a translation, reflects pretty accur-
ately the spirit of the song of Cairbre Mac Etain against Eocho
Bres Mac Elatha upon this important occasion. It is reputed
to be the first satire uttered in Erin ; and if so, is good for a
beginner. The portion not in triplets is inserted as a con-
venient introduction to the previous record of the niggardly
Alcibiades of the Tuatha de Danan, to whom he belonged on
the mother's side. Elatha his father was not a Dannite, but a
sea-king, probably in the first instance from the Spanish Bregia,
and afterwards settled in Britain. For my present purpose, as
I have represented him as looking to the gathering of the
scattered tribes, I must consider him as a Simeonite or Gaddite
by descent.

Chapter XIX

Of the deposition of Bres the son of Elatha as leader of the host,
and the appointment of Nuadh of the Silver Hand in his
stead.

Now cast I mine eyes towards Jochad who
 hearkened to Cairbre's song
In sorrow, for greatly he loved his fellow that did
 this wrong,
And therefore answered me not, nor spake when
 voices arose
Crying for him and Nuadh. Then watching these
 matters close
My God gave help. Though I yearned that
 Jochad might lead, I knew
His will was not to the spear, and only with need
 he drew
The sword from its sheath in battle. Moreover,
 meseemed that I
Was little advised of these things, lacking strength
 to descry
Wherein I might choose ; and therefore I watched
 long time their debate,

Till it rose in stormwinds of fury and howled in
    tempests of hate.

Then shook I the chain of silence,* bidding Aci
    proclaim my peace;

And he with a voice of thunder compelled their
    strivings to cease,

And aiding the son of Eschmun set forth stones
    on the ground,

Whereon the names of the captains of all the
    hundreds were found;

Yet Jochad's was set not with them, and this was
    done by my will;

For Jochad answered my glance with a brow un-
    troubled and still.

Then the throng passed by before me, and each
    man carried a stone,

Laying it as I ordered, but choice was with him
    alone

Of the wand whereby he should cast it. The
    heap about Nuadh grew

Till it capped the name which was written, but
    the castings for Bres were few,

---

* Hung by the side of the monarchs, and probably orna-
mented with small bells.

And Ogma Ethdan and Aci had each a mound to
    his name,

And stones were given by some unto champions
    of lesser fame ;

But Crimthann plucked forth his staff, nor would
    he cast his stone,

Saying he loved not to lead another band than
    his own ;

And Balor's men were away ; therefore his lot
    was bare,

And the Breogan down in the South in that
    council had scorned to share,

Saying they held their coasts, and payed neither
    tax nor tythe,

Having armour and spears for all men, and hoping
    therewith to thrive ;

So their princes came not to Crofinn. Little
    need was to count

The stones, but the son of Eschmun reckoned a
    sure amount,

Four hundred and six unto Nuadh, to Bres but
    fifty and three.

Then darkness fell upon Bres, and fiercely he
    cried on me

"Thou shalt dearly rue thy castings," and in
    answer I was not slack.

"The queen casts lots for no man." But the
    cloud hung heavy and black

As he turned to his booth and left us, and Jochad
    my husband went

And reasoned therein, but left him in silence and
    ill content,

And that night he rode to Pen Edair; and this
    was beginning of all

The strife that arose thereafter, and of many a
    brave man's fall.

Yet my soul rejoiced over Nuadh, to witness the
    patient man

Who braved wounds and neglect in silence ride
    forth at head of his clan,

Waving his keen bright spear aloft in one shining
    hand,

And bearing high in the other the mace of his old
    command

Amidst the shouts of the miledh; and he rode
    by my seat to cry

"O, queen, we are thine for ever. We die in thy
    name, Tephi."

Then my heart rose up as a queen's, and I spake,
"Nay, not with the rod,

Or the spear will I rule this island, but reign in
the strength of God."

Oh, mad are my people's shoutings.   Their hearts
are carried away.

In love of my folk thenceforward I travail both
night and day.

CHAPTER XX

(1) *Tephi goeth to the North to behold her land, and Ethan part-*
ing from her train is taken captive by Tethra and certain
firbolgs that are with him ; (2) Jochad goeth to seek him,
*and leadeth him back to their company.**

(1) WHEN the days of assembly ended, we went unto
fair Emain

Where Nuadh entertained us, and so by river and
plain

Through the North.   A hundred chosen men as
our guards he sent,

---

* This episode took place later, after the battle of Magh
Tuireadh, and Lugaid the son of Ith was Eocaid's companion
in the rescue of Ethan, otherwise spoken of as Abchan, or
Uaithne, from Tethra and his rough followers.   But I have
killed Ethan in the battle.

And fifty warriors of Dan, who with helms to
    their horsemanes bent

And sharp stiff spears before, were strongest
    arrows of fight,

For the steeds that were under these sped each
    like a shaft in flight.

Then turned we again towards Mulach where
    Maistiu would have us stay ;

But e'en as we went from the North a little space
    on our way

A thing befell which was evil, and showed the
    wrongs of my land,

For Tethra the fomorc champion lurked with a
    savage band

Of firbolgs in hills by the sea, and nought were
    we told of this

For the coastmen helped the fomorcs, though
    knowing the farms should miss

Many sheep and oxen and swine. Now Ethan,
    going apart

To assuage his soul with silence in some sudden
    blackness of heart,

Which ofttimes came upon him and drove him
    forth to the field,

By these firbolgs was carried captive.   Sore was
      he loath to yield,
But swordless and lone on the mountain ; and all
      of us angered sore
At that word.   Then bade I our miledh to search
      the hills and restore
Our bard to our train ; but Jochad ever wary and
      brave
Said "nay, yon hills and their quagmires should
      be many a miledh's grave
Hunting these goats amongst them.   These
      shaggy firbolgs will hide,
Each with his pouch of stones at his waist on the
      mountain side,
Where the horsemen may not seek him, and the
      footman climbeth aloft
Till he comes to some mossgreen hollow where
      the footing is foul and soft.
Then cometh a stone from a crag, and its hurler
      creepeth away,
Whilst the miledh if he be scatheless is stayed by
      water and clay.
Myself shall seek after Ethan."   Then cried I
      against him ; but, still

Yet strongly, of right he spake. At the last, I
 gave him my will
That he went, though my heart was heavy. In a
 mantle of green went he,
Barefoot with his harp before him, and his garments scarce to his knee
As a harper goeth unarmoured, and therefore
 unhurt of men,
Alone in the heart of the mountains to seek these
 wolves in their den.

(2) Now Jochad had skill of their customs, and knew
 their wont was to feast
On the stolen mountain cattle, and sleep like the
 savage beast
'Neath the sky, but had meat in plenty, and song
 was sweet in their ear ;
And if these had taken Ethan, it was that they
 longed to hear
The magic of Ethan's singing, but Ethan was
 wroth and stayed
Both his tongue and harp, and sware no music of
 his should be played

Before swine ; thus the men were angry, and
    surely had sold him forth
To go as a slave with Tethra to serve some chief
    of the North.
Now their track was followed by Jochad till he
    came to a pasture wild
Where Tethra was with the firbolgs, both man
    and woman and child,
And they set their meats before him, and soon
    he arose to play,
Playing the gentraith swiftly till their heels were
    frolic and gay,
And they drank and danced to the gentraith till
    after the sun was set.
Then he changed the string of his playing, and
    the wildmen's eyes were wet
At the plaintive sorrow of goltraiths, most mourn-
    ful his harp and slow
Whilst he chanted the dirge of Clidna and many a
    tale of woe
Till the eyes of them all grew heavy, and further
    they might not weep,
So low he murmured the swantraith and soothed
    their souls into sleep ;

Then gently playing he stirred, and murmuring
 still, untied
The bonds of Ethan and left them, and played
 down the valley side
Till swift on the moor they departed, and came
 to us ere the morn,
Ethan silent and shamed, but like a thrush from
 the thorn
Was the homeward whistle of Jochad. Now all
 the hours of the night
I had sorrowed upon and blamed them, but an
 hour ere dawning of light
I heard the whistle of Jochad, and stood in the
 door of my tent
And railed at my early waking, till Ethan followed
 my bent
And we three had mirth together. Then said
 Ethan, " Queen, mistress mine,
Ye be like and unlike together, but in likeness ye
 are divine,
And holy in all unlikeness : Being pure, ye are
 merry of heart.
Ye are both too proud and humble of one that
 lacks soul to depart ;

Who is proud where ye are humble, and humbled
where ye are proud,

And pardoned, lacks grace to crawl as a worm for
a grace allowed."

CHAPTER XXI

(1) *Tephi cometh to Mulach,* * *and seeth there the evil wrought
by Grisbane, the daughter of Richis upon Maistiu, and
the slaying of her thereafter ;* (2) *She telleth the state of
Maistiu in the blindness which hath fallen upon her by
Grisbane, the Canaanitish woman.*

(1) So came we to Maistiu to Mullagh. She made
us a merry cheer.

Her brow was open and happy. Her eyes were
steadfast and clear,

Yet often they fell upon Ethan, and as she sat by
her warp †

With her needle painting blossoms she loved the
voice of the harp

On the flowery banks beside her. This thing in
mine eyes seemed good,

---

* Mulach, now Mullaghmast, *i.e.*, the wood of Maistiu.

† Maistiu was the best embroideress of Eriu, and the first who
embroidered a cross upon a garment for Angus, Tephi's second
son.

L

For many spirits had Ethan, and his was a noble
    blood

Of the princes of Dan, yet lower ; whilst Maistiu
    lofty and pure

Was a queen to rule all spirits of man from a
    height secure :

But there came a guest unto Maistiu, a Canaanite
    from the South,

Grisbane, daughter of Richis.   A poppy bloomed
    in her mouth,

Her eyes danced sapphire sparkles.   A baal-fire
    gleamed in her hair

Of ruby and gold and amber, for the woman was
    very fair,

Skilled in the twisting of tiars or stringing gems
    for the neck,

And her own was white as hawthorn.   On her
    snowy arms no speck

Was discerned on their round whiteness ; but evil
    of heart was she,

And skilled in unholy cunning, knowing the fruit
    of the tree

Which is harmful, and herbs that are deadly, and
    fashioning charms thereof

To slay the spirit of man or kindle his soul to
love.

Long time was this witch betrothed unto Bennan
the son of Kain,*

But chose for her sport to tarry, and still unwed to
remain,

Casting her nets on champions. Upon Ethan
now was her cast,

With spells to draw him beside her. Therefore
it pleased her at last

To send him a tryst in the beechwood; yet, I
know not if he were weak

And minded to Grisbane's kisses, but she doubted
not he would seek

Her tryst, and herself went thither. Now chanced
it by luckless hap

I was weary within that even, and cast my shreds
from my lap

Whereon had been Maistiu's lessons, and called
her forth to the wood

---

* I have taken a license here. Bennan does not enter this tale
at all, whilst the man beloved of Maistiu and Grisbane with such
tragic results was Daire, son of Eocho Taebfhada, for whom I
have no use elsewhere.

Where she walked in her height beside me until
in a path we stood

Of soft grass amidst the hazels. There I was
minded to stay

Whilst Maistiu plucking the filberts slowly went
on her way

Down the green glade before me most lovely and
tall and fair,

With all the flame of the sunset alight in her
golden hair,

When I hear a voice beside her, "My love thou
art come full late,"

Then a sudden cry and a speech upraised in
anger and hate,

"He sends Bennan's leman to mock me, but
ne'er shalt thou mock again.

Who mocketh at Richis' daughter hath blindness,
foulness and pain."

Then one screamed, and I ran in terror, and low
on the mossy ground

Lay Maistiu, lay my sister, but blemish of blood
was not found

Upon her, though deathly anguish furrowed the
broad white brow

And a darkened juice oozed slowly 'twixt the
    close-shut lids below

Wherewith the skin was purpled.　So sank I down
    at the spot

Deeming her slain, but she moved and said to
    me, " Touch me not

Lest the poison work upon thee.　Bring water,"
    she whispered low,

And my mind flew swift in circles, debating
    hither and fro

To stay or leave her defenceless, but quickly I
    kissed her lips,

And praying quitted her side, to slip as a fawn
    that slips

Through the brake till I found the open, and
    chanced upon Ethan near,

Who [free and glad at a mark was tossing his
    hunting spear.

Swiftly I told our hap and returned.　As a hound
    that flees

At the stag, sped Ethan for water, and found us,
    and on his knees

He bathed the poison from Maistiu in silence.
    A woman's skill

Was in the fingers of Ethan, yet I feared that the
hurt should kill,

For Maistiu spake not and stirred not, nor might
we move her to quaff

From the vessel of clear spring water.   Then
was a mocking laugh

Beside us. "Never again shall thy leman behold
the day,

Or smile in thy smiles for ever.   Too skilled was
my mother's way

Of mixing her charms to fail me."   Then Ethan
rose to his feet

Raising the pitcher aloft, and hurled it down till
it beat

Full on the face of Grisbane, surely a weight like
lead,

At his knee she kneeled and stumbled.   At his
feet she fell down dead.

(2) Yes, blind, ever blind thereafter, unto the end of
her days,

Yet cheerful therewithal winning great affection
and praise.

Where she might not broider her flowers she prac-
tised a cunning craft

Of her own with a fish-hook straightened, and
  raised up her face and laughed
When I praised her taste in the colours. My
  children loved her and clung
Round her knees for kisses and stories. Many
  tears both of old and young
Water the flowers o'er Maistiu.—Of Ethan an
  eric fine
Was claimed by Richis of Breogan, a merchant
  who drew forth wine
And armour and vessels from Tarshish ; but
  message I sent him back
That Grisbane had sought her slaying, and well
  for her none was slack
To answer such woman's prayer which saved her-
  self from the stake ;
For scarce had I pardoned Grisbane even for
  Maistiu's sake,
Who prayed me towards softer answer. Our
  Ethan was soft with her
And gentle to all her teachings, but he brooked
  not any spur,
Scarcely my touch thereafter, oft hiding himself
  afar,

At times returning with songs which stirred up
    men's hearts to war,

At times returning with dirges he sang with a
    face like death,

At whiles with riddles the priesthood debated
    with angry breath.

Much did my heart lean towards him. Were I
    not set as queen

With Jochad my love, by Maistiu my chosen
    portion had been

When I saw him lying before her with the dews
    of grief in his eye.

And the Lord that knoweth the heart, hereafter
    shall tell me why.

Chapter XXII

(1) *Bres seeking aid of Elatha and finding it not, sendeth unto*
    Balor lord of the isles, and to the provinces of the north
    and the firbolgs. Crimthann undertakes to guard the
    western shore. Confusion is in the land and counsel
    undecided.

(1) Now came ill tidings to Mulach, for Bres in
    Elatha's hall

Sought aid, but his father heard him and helped
    not his son at all,

Beholding his firstborn angered, yet causeless in
    ill content.

For Bres came unto his presence, and thus their
    discourses went.

Said Elatha, "Welcome, oh Bres, but wherefore
    now art thou come

When charge of the miledhs of Eriu forbiddeth
    thee long to roam." *

"I have left them, I plundered their gold, and
    now in the mire they rout

In fury and hunger for roots, and are fain to cast
    me out."

"My son, the good of a man is naught by the
    good of a land."

"I have sucked the fruit of the soil, but fain
    again would I stand

On the necks of the men I hated, and set their
    houses to flame."

"My son, thou speakest before me the words of
    an open shame,

Be sure of this, that a kingdom never again shall
    plight

* This conversation still exists.

To an unjust seeker the faith betrayed of one
    that had right."

So Bres flung out from his father and hurried
    into the north

And gathered the barks of the fomorcs that
    through all the islands go forth,

And summoned the Sgiath and Galls, and sent
    forth men to the west

Unto Balor, Indech and Bennan, with gold to
    help in the quest

Of their coastmen hillmen and fomorcs. These
    promised him certain aid,

And Corrgen only of Ailech refused the askings
    he made.

Crimthann answered him not, as always his
    custom had been

Unto men, but sent me a script wherein he
    named me as queen,

And wrote, " Thou hast builded a throne if its
    base be the noble's will,

But mind thee that over his serfs the Chief is the
    chieftain still.

Bid me to fight with a chief, I will answer then
    at thy call.

But I wrestle not with my swineherds, nor throw
    with cooks for a fall."

So I sent him a message back, " To the queen is
    thy word made plain,

And she biddeth thee keep thy house against
    king-thieves of the main,

Which is no ill service to Eriu, nor unbefitting a
    chief."

Then came a captain of his from his keep with
    an answer brief,

" I obey," and Jochad approved me; but chiefly
    he set his care

On Bregia. Before this day the Breogan had
    little share

In the deeds of the regions northwards. Strong
    were their men and tall,

Their weapons mighty and many, their cashels
    fenced with a wall,

Whilst their traders rich within them drew to-
    gether as one.

Now Jochad feared that in Grsibane the hope of
    their peace was gone.

If their spears were against us Nuadh should be
    but a feeble strength ;

Therefore we called him from Emain and heard
    these matters at length ;

And he spake of his miledh unpaid, save his own
    band the most were lax

To practice, and many escaped ; whilst Bres had
    handled the tax

Witholding their food and armour, and now few
    taxes were paid

For the miledh, but many to Baal, the people
    waxing afraid

At cursings of priests, and rumours of war ; yet
    the tax of gold

Was paid to the fomorcs, but failed their thievish
    vessels to hold.

These had harried the coast of the north, and
    pillaged the island of Mod.

Where they burned the house of Ogma, and beat
    his men with a rod,

Whilst they set them to bind his timbers fair into
    many a raft,

And bore them away to Lochlann each at the
    heels of his craft.

Nuadh, though fieryhearted, told us no braggart's
    lies.

He longed as a steed for battle, but yet was wary
    and wise.

Braggarts came thither to us, and most of the
    common folk

And farmers believed that I by spells might
    lighten their yoke.

I know that the Lord is mighty with little or
    great to find

An aid, but as queen mine office was all my
    people to bind

In one, not kindle their strifes; so leaned I on
    Nuadh's word

And on Sri and my husband Jochad, and sware
    I would lift no sword

If other resource there might be.   Much weighty
    discourse we had.

The land being vexed with tumult, the hearts of
    the rulers bad.

Now mostly we feared that Breogan might set
    themselves to our harm,

Then said I before them all, "I have neither
    spells nor a charm

To blast like the witches of Breogan ; yet ye have
    heard the fall

Of Ai. If God be with us, the shields of the
coastmen's wall

Shall fail at my word. Then Jochad and Sri
beheld me and saw

How my heart had hidden purpose, and my will
unto these was law.

CHAPTER XXIII

(1) *Dala scorneth in the gate of Mulach, and is discomfited by
Ethan;* (2) *Tephi goeth to his relief, and meeteth Lugaid
the son of Ith of the Breogan, who was come out against
her;* (3) *she leadeth Lugaid unto her husband, having the
most part of the Breogan with her.*

(1) NEXT morn departed Nuadh to summon the
chiefs of the host

To Emain, and nigh to our gate came a heathen
bard with a boast

How Balor was drawn unto Bres, and those
would make me a feast

Unto every unclean bird and to every noisome
beast;

And my miledh were little to peck at though few
should be left alive

" The horses of Balor a thousand, his chariots one
  hundred and five,

The men of his hills five thousand, four from his
  septs in the plain.

Of the miledh of Bregin three thousand draw
  nigh from the southern main,

And Crimthann shall be behind thee with the
  war-wolves of Pen Edair

That are never slack to their hunting. Yea,
  surely they shall not spare."

Now, save that fighting in battle a bard is sacred
  of men,

Surely an arrow had sped from our fences and
  slain him then,

But Ethan was angered, and ran from the watch-
  gate, and cried his name,

" Ho Dala, called son of Cliath,* that knows
  not his mother's shame,

Called also son of the swineherd, called also son
  of the groom,

It seems in Carnamatirech † thou findest but
  little room.

---

* A harper of the 3rd rank.
† The fort of the wolves. ˚Still in fair preservation.

Outcast by Bennan the swine, Nay, that is a
wrong indeed.

Though he rout thee away from his trough, I
fling thee food for the need

Of thy mouth, three mouths in gaping ; of thy
teeth ill ordered but great,

That thy paunch which sags before thee may rise
up in high estate.

May it fill thy hunger, oh Dala, and stay the
edge of that note

Of famine above the hoarseness of crows which
dwells in thy throat

When thou singest the praise of Bennan."
Therewith an apple he sped

Large but of early Summer, and smote the mouth
in the head

Of Dala, the son of Cliath, and brake the half of
his teeth

Parting his jaws asunder, whilst blood ran stream-
ing beneath.

He might not answer to Ethan, but staggering,
turned him back

And shamed by scorn of our grooms with totter-
ing limbs and slack

Passed down the path to the meadows.   I heard
the sound of their cheer,

And leaving my maidens alone, to the guard at
our gate drew near,

And beholding him driven away, enquired of
wherefore he went,

And saw him fall on his face as he drew to a
broad-stretched tent

Some stranger had pitched there at morn, but
none came forth to his aid;

So I took a vessel of water, and ran, and was not
afraid.

Then Ethan and Sri ran after, but I waved them
back from the field,

And came on its sward to Dala, and down by his
corse I kneeled,

And brake the fruit from his jaws, and cleansed
them of blood, and poured

A wine of the South therein that was given by
Ith the lord

Of Tarshish, sunlight and honey.   Then after a
space he woke,

But his eyes were troubled and weary and never
a word he spoke.

M

(2) Still bathed I his front with water when I guessed
    behind me the tread
Of one that came from that tent, so pausing I
    raised my head
And saw one mighty of stature, the plates of
    whose greaves were gilt,
The sheath of whose sword shone rubies, and
    hung from a golden hilt,
The breadth of whose breast was spacious, and
    scaled with an armour of gold,
Dark bearded, yet white and ruddy, with features
    of princely mould ;
And he spake, " Do elves of Eriu go forth in her
    fields by day
To work their charms, and draw the soul from
    the lips and slay ?
So would I be slain if thou willest, but what is
    that potent charm
Wherewith thou hast restored him ? Wouldst
    thou work him a further harm ? "
Then smiling I said, "No charm, but wine I
    poured in his mouth
To help him out of his swoon. In vines of the
    warmer South

Was it grown of the best of the land, for in
    Gadesh the men of Ith
The lord of Breogan and Eber have vines and
    are rich therewith."
Then that mighty chief was stirred, and took my
    phial to his hand
And said, "Yea, this is of Gadesh, what knowest
    thou of that land,
If woman not spirit thou art? for never such
    sight, I ween,
Before the tent of Lugaid as thee and thy garb
    was seen."
Then joyous I said, "Oh Lugaid, art thou the
    son of the soul
Of him that named me his daughter, who, brook-
    ing no chief's control,
Went out with thine own five vessels to seek thee
    a home, and build
Thee a house wherein to rule. Thy father heard
    thou wast killed
On the seas, and mourned, and told me thy tale.
    Why then art thou here?
I was but his child by choice; but thou his true
    son shouldst cheer

The eyes and ears of his age."   "If thou art my
    sister," he said,
" I seem to hear and see the voice of one that is
    dead,
My mother, but set that by.   I am here to speak
    with the folk
Whom Jochad brings from the middens and
    hovels and stables and yoke,
To find there some champion.   I sailed upon
    many seas till I found
A people of Breogan.   There, I drew my ships
    to the ground
To reign as a prince amongst them, and though I
    love not the chiefs
Of the inland clans, they are fellows.   I share not
    a bard's beliefs
That men be equal, and seek to see if my equal
    they find
In Ogma, or Ethdan, slaves of the fomorcs time
    out of mind,
Or in Jochad, strong though men speak him, or
    perchance in one of his serfs
That dips in his chief's own basin a paw well
    dyed in the turfs.

Thus sped I before my Breogan, and now wilt
    thou pass with me

If thy sick man be helped, with my challenge;
    and soon forsooth thou shalt see

And praise thy brother as victor."   Then seeing
    that Dala rose

And departed, I went with Lugaid, and spake at
    his arm drawn close,

Towards the ditch we digged on the hilltop, and
    when Ethan and Sri would lay

Themselves in our path, I raised my hand till
    they went away.

Then Lugaid raised up his voice and shouted,
    "Oh, heremon,

Called from thy farmer folk, wouldst thou speak
    with a chief alone?

Some call thee a sheep-dog only, some speak thee
    a clumsy bear. *

I fain would know thee a lion, if not, flee forth
    as a hare

From Lugaid, whose spear is mighty; from
    Lugaid, whose miledh shall stand

* *Garbh*, the rugged.

As a wall of brass before thee, and break the
strength of thy band
Ere it fall to the wolves of Balor, the swine of the
central plain
And the mountain bulls that bellow with Bennan
the son of Kain."

(3) Then saw I a golden helmet gleam by our fence
of stake.
A light leap over the trench made Jochad, but
naught he spake,
Coming down the slopes to meet us, whilst I saw
the hurdles start
And tips of a score of arrows wait eager for
Lugaid's heart.
Naught but a cloudless wonder dwelt on my
husband's face,
As with words of happy greeting he came to our
resting place.
" Thou hast greeted the queen, by thine armour I
know thou hast titles and fame,
By sea and land, but neither thy father's house
or thy name.

Thou shalt be a champion cf Breogan, those
    ancient seamen and brave,

Sons of the sons of them that rule on the ocean
    wave

Far southward into the sunlands." Then spake
    I, " Lo, I am here

To bring thee my brother, Lugaid, the son of my
    father dear,

The old man I loved in Tarshish when I dwelt
    in his house awhile,

Who gave me the men that brought me unto
    thee and thy fair green isle.

Now my brother bringeth me Breogan." Then
    deep in his beard low laughed

Strong Lugaid and said, " More deadly hath been
    the magic I quaffed

Than his whose teeth had been broken. But
    now I see thee aright

For a lion, I have my longing, and hail thee a
    lord of fight

Who shall shame no man as his captain, and
    Balor is none of mine,

Though he may perchance excel me in strength
    to wrestle with wine,

And Bres may win at the chess-play. I bow to
 thy queen great righ *

And thy helm with her ruby above it. Thy man
 henceforward am I."

Then Jochad embraced him and said, " My queen,
 my mistress, my bride,

This day thou art champion of war, the chiefest
 strength of our side."

And Lugaid laughed, " It is little thy queen hath
 conquered in me ;

But the daughter of Ith may call the sons of the
 sons of the sea,

And win back a loyal answer. Fair queen, so
 haughty and small,

Say wilt thou travel with me to set on thy crown
 the wall

Of the Breogan towns of the South to keep thee
 here on thy hill."

Then Jochad was grave, but I smiled, and he
 spake not against my will

When I followed Lugaid afoot till he set me on
 Enbarr his steed

---

* *Righ*, king.

And went by my side five furlongs. Now
    whither our road should lead

I had guessed. O'er a rough rock's shoulder we
    climbed and below us stood

The miledh of Bregia camping betwixt that cliff
    and a wood.

At Lugaid's shouting they turned and knew him
    and drew anigh

Whilst he spake of me to his men, for that crag
    was set too high

For my speech to pass to their ears, but high on
    the topmost stone

I stood few paces above him, and a thought I
    had made my own

Was this. The trident of gold I had from the
    Pen of the Gate

Should be known of these with the twiceforked
    spears. By a happy fate

I had seen my maidens bearing it forth in my
    house that day,

And chosen this for a rod, and a weapon to be
    my stay

When I went down the field to Dala. Now I
    raised it on high

That its threefold fangs of gold might lighten
    against the sky ;

And the miledh hailed their standard, for many a
    grandson of Tyre

Knew in what temple shone in the god's hand
    such dart of fire,

And great was the shouting then, though some
    of the folk were wroth,

Till there came division amongst them, and part
    of their band drew forth

With Richis to go unto Balor, but more than the
    half turned back

And passed by the crag, and followed where
    Lugaid pointed their track.

Two hours had I gone from Mulach, when again
    I might discern

Once more the eyes through the wattles that
    waited on my return,

For none might pass through the trench save
    Jochad gave them command.

I that departed with one, returned with an
    armoured band,

Twelve hundred and three and fifty, whilst some
    stole thither by night

Until Breogan stood fourteen hundred, a wall to
    hearten our fight;

With Lugaid the stone of their corner, the prow
    of the thorny hedge

That should brush the horsemen asunder, as a
    swan that stirreth the sedge.

CHAPTER XXIV

(1) *Lugaid journeying with them meeteth his father by the way,
who is secretly slain by three Canaanites thereafter;* (2)
Lugaid maketh jest of the porters at the gate of Emain;
(3) *The tribute is cut off.*

(1) AT the dawn I said, "let us carry to Nuadh the
    Breogan aid,

That his soul be uplifted with us, and his miledh
    be not dismayed

By tidings both North and South. So I and my
    husband led

With Lugaid, and Ogma tarried a space behind
    at the head

Of our folk and the men of Bregia. Then, pass-
    ing on without fear

We saw on our path a greybeard most noble of
    horse and gear

Who came in the way before us ; and now, behold, it was Ith,

And he fell on the neck of Lugaid, and great was
our joy therewith.

Beholding his son he wept ; and gave to the Lord
great praise

That his eyes found light to behold him, before
the darkness of days.

Tidings had come out of Bregia that his son was
living as yet,

Thereupon he made no tarrying, but quickly his
course was set

To see if that word were true ; and now, than his
hope more swift,

His son had kneeled for his pardon.   Then both
did their gaze uplift

To my face, and he kissed me also, and blessed
me of heaven that his son

Was found, and had counsel by me, and bade
him his course to run

'Neath the eyes of his daughter Tephi, enquiring
much of our war.

Then said he, " Ye call me, Ith Cian, the ' light
that liveth afar,'

In this land where my ships come often, but soon
    shall ye see me near.
I am not too weak in mine age to handle the
    sword and spear.
I speed and return with succours. One hour
    with ye I remain :
Then back unto Edair's harbour to summons the
    ships of Spain.
In a month hence abide my coming. My going
    shall not be long.
My ships shall be very many, their engines and
    armour strong."
He heeded not for our chiding. "Nay, I have
    seen my son
My very son, Lugaid, in right. My journey is
    wellnigh run.
Let me strike one stroke against Balor. He also
    is mighty, yet old.
His seawolves have oft sped southwards to harry
    sheep of my fold."
Thus spake he, and would not tarry ; yet scarce
    had he left our sight,
Riding full swiftly to Edair, when now at entrance
    of night

Three champions of Tyre drew nigh, and though
    the even was dim
They guessed of Ith by his riding, and their rid-
    ing was known unto him,
For he drove them forth out of Eber, being proud
    that no man might stand
Of the chiefs of Eber before them, and haughty
    in all the land;
Yet valiant and strong and wealthy. Now these
    were sworn unto hate
Of the lord of Tarshish, therefore he turned him-
    self by the gate
Of a farmstead amongst the cattle, but the eldest
    man of the three
Beheld him and followed after, and beat him
    down on his knee
Whilst his brothers slew him with stones, and
    after they builded a heap
Of the stones above Ith Cian, and trusted their
    deed would sleep:
But ye know, and therefore I write not, the tale
    that the bards shall tell
To the sons of men for ever, how these princes
    of Canaan fell

'Neath the burdens of Lugaid upon them.
Though greatly they strove therewith,
They were laid at the last 'neath the stones where-
under they buried Ith.
We knew not this on that night, yet deemed that
Ith was no more
When his succours came not from Tarshish,
knowing the love he bore
To his daughter and son, and his wrath against
Balor, Indech and Bres.

(2) Yet this night we guessed not his doom, and went
without heaviness ;
And the next day drew unto Emain, riding
thither full fast
Before our people, and Lugaid swore that a jest
to last
Should be in our coming thither. So went he
afoot to the hall,
His brightness veiled by a cloak. Now there
stood two guardians tall
And haughty by Nuadh's threshold, and these
men bade him to stay

Until his errand was told them.　Then said he
　　humbly, "I pray,
Doth Nuadh require a wheelwright?" and the
　　porters answered him "Nay,
We have Luchta, the son of Lomhaid."　Then
　　asked he again, "I pray
Your favour, wants he a smith," and the porters
　　again said "Nay,
Our smith is the thrice-skilled Colum."　Then
　　bolder he spoke, "I pray
Lack ye here for a champion?" and loudly the
　　men cried "Nay,
Great Ogma cometh and Ethdan."　Then sweetly
　　he sung, "I pray,
Want ye my songs as a harper?" and proudly
　　they answered "Nay,
For Ethan comes oft to our tables."　So, solemn,
　　he asked, "I pray,
Have ye preachers and pious amongst you,"
　　and scornful they spake him, "Ay,
The wisdom of Sri, the preaching of Mathgen."
　　So laughed he, "I pray,
Are cupbearer's near to your lord?"　They
　　answered in mocking, "Ay,

Dathi leads twelve clad in crimson?" Then,
    formal, he questioned, "Pray,
Be there scribes or recorders with them?" Where-
    upon they answered him, "Ay,
Many scribes under En son of Eschmun." So,
    last he said, "I beseech
Your mercy in asking, hath Nuadh provided a
    skilful leech?"
One laughed and the other yawned. "The chief
    of that craft have we,
With son and daughter beside him, wellnigh as
    skilful as he."
Then Lugaid cast cloak, and shouted, "Go,
    Kamal the son of Knees
And Hamal son of Formality, ask thy master, of
    these
Which man may do every service?" Right
    swiftly these lackeys sped
At his chiding, and Nuadh heard them, and came
    to the gate and led
The "man of all crafts" * to his table, where
    laughter and mirth we found

* "Ildanach," a title of Lugaid's, who may have picked up
his oriental terms of abuse (Gamul Mac Figol and Chamal Mac

To greet us upon our coming, whilst gaily that
    jest went round.

(3) Now as we sat at our meat, there came nine
    men with demand
That the tributes set by the fomorcs be given into
    their hand ;
And spake with threats in their mouths that the
    taxings be swiftly made,
Bidding us hear that thereafter a double tax
    should be paid,
If Balor and Tethra should tarry, or Indech
    should stay his oars
That he sent unto Losken-lomu, to bring with
    speed to our shores
His barekneed kernes from the North. Then
    stood I before these men
And said, "The Shepherd of Israel keepeth
    wolves from the pen,

Rhiagild) in his wanderings, or learned them of the folk whom
he is reported to have sent as far as the Persian court for steel
weapons, probably unobtainable further west at that period.
The physician's name was Diancecht, the lady doctor's
Armedda.

His flock shall be tythed of no man." Then
    Lugaid arose in wrath
And falling swift on the seafolk, with the spear-
    staff he drove them forth,
To return unto Indech and Balor. But all hearts
    gathered to me,
For my labour was fallen upon me, and my
    travail for victory.

Chapter XXV

(1) *Tephi holdeth her council at Grellach Dollaid, and cheereth
the men of Eriu ; (2) Eocaid gathers his force of the men
of the land and of the horsemen of Dan, whilst Lugaid
goeth to the South and Ogma to the North. They make
their trysting in the West, by the water which is now
called Unius, and Tephi sendeth messages to Elatha.*

(1) Old Nuadh's heart rose up as a man of war to
    cheer
  Our hearts, a steed that snuffeth and knoweth
    the battle near,
  And we planned our secret council that was held
    on a Sabbath day,
  For our righteousness is with the Lord in our
    toiling as when we pray.

In a hidden hold we made it, of the chosen of
    all our land,

And greatly the people marvelled of the deed
    which thereat was planned,

Wherefore men call it my amrun,* for all men
    marvelled to see

How God spake forth in Eriu by the Spirit He
    set on me.

Now after a while, I bade that each man speak of
    the gift

He would give unto God and Eriu the burdens
    thereof to uplift,

Then Mathgen the wise said, " I and the priests
    through the hills seek aid,"

And Figol son of Manoah, " Oft on my knees I
    have prayed

Amongst the men of the woodlands, and surely
    these know me well,

And will seek at my bidding to Tephi to fight
    with the powers of hell."

Bright Dathi said, " I am known by many a river
    and lake

---

* A marvel.

To the aire's and shepherds, and these will surely
    come for my sake."

And Lugaid, "Of Breogan, my strength, I issue
    forth with my spear,

The Destroyer, with Perez the Mede its light-
    nings were seen with fear.

None such hath been known in Eriu. 'Tis a
    flame of thrice-tempered steel."

Now many spake of their will for the good of the
    land to deal.

Gabhran the smith saying, "Never shall freedman
    of Eriu want

For spearheads or bolts or javelins till the coals
    of my forge be scant."

And Luchtna, "For Gabhran's spearheads such
    shafts will I surely make,

As shall fill each outstretched hand, and no one
    of my shafts shall break."

And Creidne, "Of every spear which Gabhran
    and Luchtua's skill

Shall fashion, the heads shall cleave, for my
    rivetting is not ill."

Last, Jochad said, "Ye have promised each and
    all as a King

Yet myself is the Queen's first servant, and there-
    fore myself I bring."
Then Lugaid smiled and he said, "The serfdom
    of all is seen
In their mouths, but what wage for labour shall
    be to thy slaves, Oh, Queen ? "
Then answered I at that asking, "Little my need
    of a slave,
But free service to this my kingdom." And
    thereon I made them a stave.

Not upon slaves are my gifts poured out.
Strong olive, anointed and diggèd about,
Mine oils are sovran o'er weakness and
    doubt.*

(2) We determined that Lugaid should pass with his
    Breogan homeward and west

---

* Arrosisor dosifius
    Dosseladh arosel
    Arrosdibu nosriast
For the difficulty of translating the Great Queen's utterances
see Whitley Stoke's " *Revue Celtique.*" I am no scholar.

And Jochad be with me at Tailtea,* whereto I
    should gather the quest

Of all the lands of my province, and also through-
    out the soil

Of Eriu send men to gather hills fields and
    pastures from toil,

Loyal folk but skilless in warfare. Yet Jochad
    had heed of all,

And taught them and gave them arms ; and their
    women and babes would fall

At my feet, and pray me to lift the curse of the
    robber bands

That issued out of the cashels, and harried the
    farmers lands

Till they lacked the oxen to plough with, and
    often they failed to eat

The very seed they had planted, for oft these
    carried the wheat.

In my tears I promised their asking, and gave
    them of that I had,

---

* The seat of Tephi in her immediate domain of Teffia
(Tephi's land), where she probably died, being carried thence
Teamuir for burial. Teffia included Longford and West-
meath.

Grown little now by my spendings, but the souls
    of my poor were glad,
Till some called me not "Teia" but "Dea," and
    save that they dwelt with the clods
I had needs reproved them more sharply, for I
    love not that names of gods
Be given to men; and after, such rebuke was
    often my need
In chiding this foolish people, but my preaching
    hath little heed.
Ogma went from us northeast, and passing a
    space inland
He drew us a noble succour of men of war to his
    band,
And passed unto Ailech to Corrgenn, and thus
    in a six weeks' space
We had gathered Eriu amongst us, and drew
    towards the trysting-place.
Where Balor and Bres should find us, and where
    should be held that fight
Which should darken the clouds of Eriu or fill its
    dwellings with light.
One thing unknown of my husband I did, for I
    feared to fall

Therein. We heard how a bridge betwixt the
    isles of the Gall *

And Eriu was wellnigh built by boats going
    hither and fro

With Sgiaths and Firbolgs in thousands, for
    Indech had not been slow

Of help unto Bres, nor Tethra, nor Omna nor
    Bagma the chiefs

Of the fomorcs, to bring with ships these bands
    of savage reliefs

Unto Balor. Then sent I word to Elatha the
    father of Bres

That the host of his son grew mighty. His
    honour grew less and less,

Bringing wild Firbolgs to plunder a kingdom
    which once his arm

Was strong to defend against them. So I told
    my husband my charm

Had been woven to weaken Indech, and surely
    my soul spake true,

For Elatha sent many vessels to harass that
    pirate crew,

* Foreigner.

And the isles of the Sgiath's and Firbolgs, till
  lastly these feared to come,
Whilst many that came already went back to
  defend their home.

CHAPTER XXVI

(1) *Tephi and her husband come to the ford of Unna * where
  Eocaid dreameth a dream which she may not interpret,
  though she is cheered thereby ; (2) the chiefs of the host
  assemble thither, and a camp is pitched, whilst the battle
  is set for the eve of Samhain † ; (3) the fighting of the first
  day, Ruadan, being treacherous, is slain by Gabhran the
  smith.*

WE were first, one week ere Samhain in the trysting by Unna's stream.
In the early dawn thereafter, my husband told
  me his dream
How I stood o'er the pool of Unna one foot on
  his own green land,
But the other firm on a lion that slept on a fair
  bright strand.
Nine braids of my locks spread forth, and lo, the
  first of a three

* " Destruction,'' named after the battle.
† October 30th.

Was wavy and many tangled in all the isles of
    the sea.

Now the second was thick and braided on a
    broad land wealthy and fair

In the West, but that tress was severed, and cities
    grew from each hair

That lay on that noble pasture. Then the third
    tress spread to the north

In a great land buried in snows, which melted till
    streams gushed forth

Amidst oceans of golden cornland. Then he
    spake of the second three,

How a thin hair, strongly braided, upheld the
    weight of the sea,

And a second stirred by a westwind flew to a
    golden hill.

Whilst its fellow gave shelter from heat o'er realms
    stretching beyond it still.

Of the third three, all went south, and one was
    spread over Lud

And Phut, but the other twain flew out o'er an
    endless flood

Unto the endings of earth, and there they fastened
    their hold

Upon mighty desert places in the heart of whose
    stones was gold.

Now on every tress of the nine were golden cym-
    bals which spoke

In the ears of the lion's cubs which lay at my
    foot : but he woke

Ere ever his dream was ended. Yet he watched
    four eagles draw

Towards the lion to blind his eyeballs, and the
    lion opened his maw

And roared in face of the eagles. Then started
    he full awake.

That dream might I ne'er interpret, yet my soul
    is glad for its sake.

(2) Yet the roaring was of young lions, for Lugaid
    and Ogma were there

With their force before the daybreak, and surely
    they did not spare

To roar as lions in their coming. Thus was our
    host complete,

And Nuadh went forth before us, and ordered a
    battle seat

On the green slope stretched before us. Noble
    was now that host,
And valiant, but little of number before the chiefs
    of the coast,
With their swarming Firbolgs and shipmen.
    Now each side ordered its fence,
And we parleyed, and set the battle of the forces
    for five days thence.
Upon Samhain's day which they chose, for this
    was a feast unto Baal,
But my Stone of defence was sure. His pillars
    of little avail.

(3) Now the plain by the stream of Unna was level
    and broad and green
Till the rising fences of Balor on a further hill
    might be seen
Whence shoutings came to our ears, and cham-
    pions out of his side
Came forth in the field and mocked us, and I
    would not any replied.
Yet often they went; and some were victors, and
    some men fell.

I might scarce forbid such strivings; but this
    thing I knew right well,
That such are not for a leader in whom a nation
    is lost,
So laid my gesa * on Lugaid and Ethdan at
    every cost
To bide in their booths with Jochad. Nuadh
    secure might ride,
For the chief of a host is sacred till his battle be
    ordered wide.
That first day were many combats of lesser men,
    and a car
Of Ochtriall son of Indech we took with his
    craisechs † of war,
When he went to stop the springs to our front,
    for the streamlet ran
Too near to their slings for our sutlers. Also
    division began
Of these, and the spears which Gabhran and
    Creidne and Luchtna made,

* *Gesa*, command with curse for disobedience.
† Craisech, a broad heavy spear with a blunt point, used by
Firbolgs and seamen.

Each with its well-poised shaft, and rivets, and
    bright keen blade,
Till the foe had heed of that forest, and at even,
    one that we knew
Came from them and went amongst us, for the
    stream of his life he drew
From a captain of Dan, though his mother was
    even a Canaanite,
In whom a chief of the fomorcs long time had his
    heart's delight.
Ruadan was his name, and much he enquired of
    our gears,
And saw where Gabhran the smith was casting
    the ruddy spears,
And Creidne plying his hammers, and Luchtna
    shaping the wood,
The three great craftsmen of Eriu, and the work
    of their hands right good
And speedy ; whilst Tuirbhi, crippled, wrought at
    his forges ill,
Though had he been strong in his prime, our
    Gabhran, his pupil, still
Was his master in skill and swiftness. Then the
    spy to Tuirbhi went back,

And told him we cast ten spears unto one, and
　　his arm was slack ;

So Ochtriall, grieved for his craisechs, moved him
　　to seek our camp,

And find if sods might be gathered the fires of
　　our forge to damp ;

And he took a spear of a woman who ground it
　　upon a wheel,

And hurled it swiftly on Gabhran, thinking there-
　　by to steal

Supply of our weapons from us; but the spear
　　that went by his back

Tore but the flesh of the smith, so Gabhran sped
　　on his track,

Drawing the head from his side, and hurled an
　　avenging stroke.

May all traitor's perish like Ruadan, whose breast-
　　bone and back were broke.

Chapter XXVII

(1) *Ogma fighting with Tethra wins Ormai his sword;* (2)
*Bres cometh to chide, and seeking Lugaid, is fought with
by Ethan the poet, who is shamefully smitten by Bennan,
the son of Kain, whom Aci, son of Alghuba, beareth dead
unto Tephi.*

(1) TETHRA, the sea-king, came next day in the midst
to deride
Both Jochad and Ethdan, but Ogma went for
them on our side,
Falling swiftly upon him, and beat him back to
their fence again.
Had Tethra not fled from Ogma, surely he then
were slain,
Having lost his sword behind him. That sword
was heavy and keen,
Its hilts well guarded, and Ogma bore it back to
the queen,
Saying, "Ormai, its name is well known." Now
graved on the blade were lines
Straight, or sloped in their groupings; therefore
I asked their designs.

o

Then Ogma said, "These be names of champions
    that Tethra slew
With Ormai in former days, and each is a record true
Of the sixteen feats that be graven." Sri also
    approved him of this,
Reading forth the champion's titles. Then out-
    spoke Ogma, "I wis,
It is well that a name remains of a miledh and of
    his deed.
If I fall, no man shall know my resting save such
    a screed
Be set on the stone that marks me." Surely so
    it was done
With grief on the headstone of Ogma that day
    when our fight was won.

(2) On the first three days flowed balsams, on the
    fourth a river of grief.
Out of their gate at morning shone bright the
    arms of a chief
Which blazed in the Autumn sunrise. A figure
    of princely mould,
Whose spears were iron of Tarshish, his buckler
    of beaten gold,

And his helmet and breastplate likewise. Then
    all men knew him for Bres,

Who came before us and spake, and his words
    were of bitterness.

"How long did I herd the swine, that now
    amongst wolves am found,

Whilst the swineherd Nuadh lay sick, when Ogma
    crouched like a hound

For my scraps, and Jochad was mine ere ever he
    gave his heart

Unto piglings routing for roots, and a woman
    bade me depart.

With none of these will I fight, for these were
    my servants all;

But lo, I behold with swineherds a champion
    slender and tall,

And meseems, well skilled in his saddle, who ne'er
    hath been dog of mine.

I will fight with him if he listeth, and the light of
    his courage shine

As bright as doth Canbarr his helmet." Then
    Lugaid grew mad for fight,

Till I angered and claimed my gesa, his champions
    holding him tight,

Yubor, Seibar, and Eru, whilst they bade him remember Spain

And the oath he made to his sire, and how he
   had right to reign

If his father indeed had perished. Still, sore was
   his mood to go

Till in the midst of our chiding, we heard a
   murmur run low

Of wonderment round our trenches, and setting
   mine eyes to the fence

I beheld how Ethan the poet like an arrow of
   war sped thence,

With shaft and sword, but unarmoured, whilst
   Bres in the open field

Laid low his spear for encounter, and eyed him
   above his shield.

Now the shaft which Ethan carried was heavy
   and sharp and thick.

Through the golden shield he hurled it, and
   leaping thereafter quick

On the spearshaft bore the shield to the ground
   with his proper weight,

And saving that Bres fell with it, surely then had
   his fate

Been death by the hand of Ethan, and Jochad
    cried, " 'Tis a feat

Most worthy a great war champion," and Lugaid
    answered in heat

"Such feat had never been mine. Nay, I knew
    not this of my sires."

Whilst Ethan smote with the sword on the helm
    with its jewelled fires

Which gleamed on the sward beneath him and
    shore away half its crest,

Then raising his hand again he smote it against
    the breast

Wounding above the mantle, but his blade on the
    buckle broke ;

Whilst Bres, being mighty, arose, and struck him
    down with the stroke

Of his spearshaft laid to the neck, whilst we
    shuddered as Ethan fell ;

But Bres set his shield above him, and we trusted
    all should be well,

When Bennan, that came by stealth from their
    fences to watch that strife,

Thrust under the shield his spear. Then Ethan,
    leaving his life,

Set eyes on Bennan and knew him and said,
  "With me there is bliss,

But the giver thereof I bless not, for love was not
  in thy kiss."

Thus died he, and Bres was moody in shame, but
  naught he spake

Striding in wrath from Bennan. Then, for God
  and my kingdom's sake,

I bade Aci son of Alghuba go swift to the son of Kain,

And command him into my judgment, and swiftly
  return again.

He ran, and he came on Bennan, and caught him
  round by the waist

Lifting him high though he fought in the arms
  which his girth enlaced

Until Aci strode in our trenches. No blood in
  that strife was shed,

But ere Buman was thrown before me, the soul
  from his black lips fled,

And he went to the Lord of Judgments. Aci
  returned with his corse,

Having message from God and his queen, he
  wrought it with mighty force.

Oh great was our mourning for Ethan, but holy
  our joy likewise.

We laid on his brow in the sidhe a champion's
   helm as his prize,

Whose badge was my spray of Olive.   There
   they dwell with his dust

Beside the waters of Unna, but his glory shall
   never rust.

CHAPTER XXVIII

(1) *Nuadh leads his forces in three bands against Balor of the
Mighty Blows, and Lugaid doth many deeds of valour in
the centre of the fight ;* (2) *The miledh upon the right
are harassed, and Nuadh trusting to slay Balor with his
darts ; is slain by him.   Indech presses sore upon the
miledh until Ogma and Indech fall by each other's spears.
Lugaid comes from the centre and slays Balor, retriev-
ing the battle of the miledh ;* (3) *Tephi watches the fighting of Jochad and the men of the land who are victorious
against the Firbolgs and Canaanites ;* (4) *The Queen
gives pardon to Bres and Tethra at their fences, and the
slain of Balor are counted by Uan Cendach his scribe ;*
(5) *Tephi maketh a song of instruction for the priests to
sing to the people.*

(1) ON that day we arose ere dawn, and the heaven
   was black with cloud

As we mustered our men on the hillslope, but of
   surety my heart was proud

Whilst they sung the warsong I made them.
"The Kings arise unto fight." *

Marching so strongly and proudly mine eyes
grew wet with the sight ;

For the most part had been but yeomen and
herdsmen out of the field,

Not men of war from their youth, nor feared I
that such would yield

To the knives and stones of the Sgiaths, but
dreaded the long-stretched wall

Of the coastfolk guarded in armour, and the force
of the men not small.

For their Firbolgs, I feared them little. The
horsemen of Dan should sweep

From our flank and ride amongst them, and slay
and drive them like sheep,

And the plain was too rough and soft for chariots.
I recked not of these,

But their strength with Balor and Indech and
Bres and the men of the seas

In three lines like a thorny fence. The first,
low couched to his shield

---

* Afraigid rig don cath. This warsong of Tephi's still exists,
but I have been unable to meet with a translation.

Till a rampart of bronze and hides stretched end-
    less across the field

With strong thorns of death before it, whilst they
    that behind it stood

Bare javelins very many which sprouted thick as
    a wood.

Upon these were cords of leather to the end that
    being cast

They are not lost in the hurling but unto the
    wrist bound fast,

To be drawn again to the seafolk. Lastly, with
    slings and darts

Stood their slaves to aid their forefront. So now
    with the thought that starts

Unbid to the lips, I ordered my Breogan to
    shorten the line,

But the fourth of our foes already, till the ranks
    of their men were nine,

And break them upon the centre. This Nuadh
    and Lugaid approved,

As Nuadh rode out to the right, and down on
    their left-hand moved

With the horse of Dan and his miledh. The left
    was my husband's place

With the multitude of our people, to carry them
face to face

Through the swarming Sgiaths and Firbolgs,
before Breogan upon their right.

Right royal he rode with his people, and cheered
their hearts for the fight.

At the centre Lugaid rode round his column his
spear in his hand

Singing " Arotroi cath comartan." * Then hurling his ninefold band

On their triple line it parted. So scattered their
swarm and brake

In surges upon his phalanx, but our shield-wall
it might not shake ;

And there was Ochtriall the leader of the fomorcs
of Uan slain,

And the might of Omna and Bagna their
champions wasted in vain.

There Luad struck down Loch Lentglass a mighty
warrior in strife

Where he lay on the ground unsworded, and
Lugaid gave him his life.

---

* A song which Lugaid made against paying tribute to the
Fomorians. It still exists.

(2) But our right-hand had nowise prospered. Brave
     were the men and true
Of the miledh that followed Nuadh, but their
     ranks were wasted and few ;
Their horsemen stayed by the clayfields. Thus,
     or ever they drew anear
To the line of Balor, in places where no man
     might thrust with the spear,
Rushed Firbolgs swiftly upon them, and hurled
     forth darts and were fled ;
So that many were wounded amongst them, and
     three captains of hundreds dead,
Ere they came to the wall of Balor. Then
     Nuadh, though old, was rash,
Beholding his ancient foeman, and went out
     swiftly to dash
Upon him ere any might stay him ; so, shouting
     his name, rode in
On the line and brake it asunder, and thought
     by that deed to win
The fight against Balor and slay him, hurling
     with mighty force
The one of his spears, which wandering, pierced
     but the head of a horse

Before the chariot of Balor. Then his second
    javelin he threw.

On the brazen shield of Balor, raised slantwise,
    it glanced askew,

Smiting Cannan, brother of Bennan. Then,
    grasping strongly his last,

Rode Nuadh to strike down Balor; but even
    now as he past

One smote the heels of his horse, and rearing
    upwards it fell,

Whilst Balor forth from his chariot leapt in the
    hate of hell

With an iron craisech, and slew him. Then
    fiercely forward his men

He drave on the miledh of Eriu, who weary
    came from the fen

And, sad with the falling of Nuadh, slow and
    sullen drew back,*

Until Indech curving his men from the left-hand
    horn in attack

Beside them, many were slain; and Indech,
    passing behind,

---

* It was at this point of the fight that Tephi's sister Maacha
was slain, as mentioned before.

Drew forth in the field with hope our camp
    unguarded to find.

Therein was his greed reproved, for Ogma, with
    chosen guards

Of the Danites, was set to keep me. Moreover,
    the scribes and bards

Had each one a champion's spear. E'en the
    priests that came with us to pray,

And the cooks sang "afraigid rig don cath" on
    that mighty day ;

With neatherds, swineherds, and boys who each
    had darts in his hand.

So great had been Gabhran's zeal that these
    looked like a warrior band

Behind the stakes we had planted. Thus,
    Indech halted anear

To behold, and Ogma, the loved one of Jochad,
    couching his spear,

Rode forth with a troop against him, and Indech
    stooping his head,

Rode also, till piercing each other, those
    champions fell down dead ;

And a great cry rose from our fences; but on the
    horsemen of Dan

Rode o'er their fallen leader, and each one slew
    him a man
Of the fomorcs, and over our fence came
    trooping the carles with spears,
Till the hearts of the men of Indech being
    smitten with idle fears
They fled to their ships from the battle; yet our
    need was sore on the right,
Where the men of Dan, with the miledh, stood
    back unto back to fight
As a rock that wastes by the sea-wave, till
    bringing the central wedge
Of our fight, bright Lugaid appeared beside them
    to set the edge
Of the Breogan sword on the fomorcs, and
    sweeping as chaff their slaves,
Parted that sea which girt them as a vessel
    parteth the waves.
Then, taking a keen-edged stone, a champion
    stone, for his sling,
He sent it amidst their chariots, and smote down
    Balor their king,
For it struck and went out behind him. Then
    riding on in his wrath

He spake with his spear unto many, bidding the
    soul fly forth,
To do service still unto Balor.

(3)   Meanwhile mine only delight
And terror had been that day to gaze on our
    left-hand fight,
Where I saw the throngs go steady, with one
    crest moving o'er all,
The tallest and brightest there.   Ah me, if that
    crest shall fall!
Now, in midst of the plain, sore is that host
    beset.
The Firbolg flood is around it.   That helm is
    not stooping yet.
See, for a moment it bends.   Behold there
    cometh a troop
Barekneed.   These be Loshken's kin.   He
    rideth head of the group.
His plaid flies wide from his brooches.   He
    beareth a mighty brand.
His fosters with targes are by him to aid him on
    either hand.

Is it Aci that smiteth his fosters? I see but the
    shining crest

Stoop twice and Loshken is fallen. Deep is the
    wound in the breast

Of Loshken-lomu Mac Lomglain, who carried
    his barekneed kernes

Out of Sgiath north unto Scetna, where the
    northernmost ocean churns

Upon rocks that are white with seafowl. Now
    are the white knees spray

Before Jochad and Aci riding, and swiftly it dies
    away

As they hammer the bronze of Breogan. Behold,
    it bends with the strain.

Yea, shout with joy, it is broken. Nay, it is
    mended again.

Eriu is slow going backward, yet steady from
    rank to rank. •

There cometh a host of horsemen, and driveth
    upon the flank.

Yea, Bres with his horsemen rideth. Surely now
    shall they flee.

Let my prayer be pure with the Lord who hath
    holpen me on the sea.

Yea, though the hail pass over. Yea, though
    the billows roll,

The Lord is the Stone of my corner, the strong
    defence of my soul.

Great are their shoutings and strivings, great is
    the clashing of swords.

The heathen are mighty and many ; their leaders
    are chosen lords ;

But that helm goes hither and thither, as a fly-
    ing star o'er the strife.

It brightens the heart of our battle. It flashes
    where men yield life

For God and for Eriu and me. The grasses are
    stained with gore,

But that heaving ceases. Oh sternly doth Eriu
    flow once more

Against the bulwarks of Breogan. Lesser is now
    their band,

Yet more swift and fierce than aforetime. Who
    at this hour may withstand

These trusting in God and their captain, these
    lifting a crushing wrong

Which bowed the necks of their fathers. Needs
    must that their will be strong

To buy with their blood this battle. Here Richis,
    the proud man, fights,
By Tuiren the son of Malek. The lofty, my
    champion smites ;
And Tuiren is slain by Aci ; but the horsemen
    again draw near.
By the left they pass behind us, and now they
    ride on the rear.
Scarce do they smite our hindmost, ere Ethdan
    cometh at speed
With horsemen of Dan behind him. He helpeth
    our sorest need.
They be many, and Dan but few, yet Dan hath
    made him a track
Betwixt the foe and our footmen. No one of my
    own turns back
To look on Sodom behind him. Each presses
    on to the mark
Where the gleaming golden helmet is set as a
    guiding spark.

(4) It is even, lo, they are yielding. Yea, they have
    called me a witch ;
But I know the distant slaughter. I hear their
    cries in the ditch

That lieth before their fences. My soul may no
    longer stay.
I mount the white steed of Jochad. Full swiftly
    I ride away
With tears and blessings behind me. Now Jochad
    and Lugaid form
Their force to a single band in the field for the
    final storm,
As I find the son of Alghuba, and bid him pro-
    claim that now
The Queen brings word from the Lord that all
    who have need shall bow
Before her and take her ransoms. This message
    therefore he cried ;
But over the speartopped fence no voice of a
    man replied.
Then, knowing many should fail ere ever its fruits
    were won,
And grieved in my heart thereof, I carry my horse
    alone
Nigh up to the trench and speak, and awe is on
    those within
From the Lord, for they deem that I alone in His
    strength shall win

The gates of their fence, so they hear, and these
were the words that I said.

"Is there any wounded within? Is there any
man sore bested?

I have leeches to tend his hurts. I have succours
to help his heart.

Moreover, if any would go, I give him grace to depart

Unharmed if he go in peace to his land ; or, if of
mine own,

I bid him kneel unto David, and seek his grace
of my throne."

Then heard I voices within, and after a space
spake Bres.

"Oh Queen, which lot were my portion? I
would not add less to less,

But more unto more. As yet, my spearmen are
more than thine.

We have strength in our fence. On our spears
the sun with the morn shall shine.

Yet, if thou holdest thy word, I promise that
never more

Shall the taxings made for the miledh go forth
from thine island shore."

" Is this the gift of a champion that would not
grow less and less ? "

I said, "Such gifts, not his own, shall not be
worthy of Bres.

Go seek Elatha, thy father.   Go spend the rest of
thy days

In ridding the seas of robbers.   Thus win thee
a champion's praise,

That thy name be increased with blessing, and
sink no more 'neath a curse.

There be good and evil before thee.   Why set
thy hand to the worse?"

Then Tethra chided with Bres, and said "We be
overthrown.

Why should we longer bide?   The half of my
men are flown,

And Tuirbhi our smith is wounded.   Let us take
the message she gives.

Now Balor and Indech are slain, what man should
vouch for our lives?

Whilst small hope is ours of a booty."   Yet think
I be moved not Bres,

For he answered to me alone.   "Behold, I am
less and less,

Yet fain would be more and more.   Therefore,
oh Queen, I will go

In the name of thy Stone hereafter ; seeking thy
grace with woe

For all I have sought with evil." Then said I
" Peace unto thee,

That the blessings of wise Elatha shall rest betwixt thee and me."

Then back ride I to my folk whilst swiftly the sky
grew gray,

Bidding all return to the fence, where I sank at
close of that day,

Being faint, but thankful of heart; and none
enquired of my deed,

Yet men of the fornorcs told it, and mighty then
was the meed

Of my praise, though some of the miledh fain had
plundered the foe,

And murmured that after his binding, I loosed
him and let him go.

Yet our spoils were great in the field, for Uan
Cendach, their scribe,

Came forth at the morn, and he named us the
names out of every tribe,

Of kings and chiefs that had fallen. Of kings
were forty and two,

And of chief men very many, whilst these on our
    side were few,

Save that Nuadh and Ogma lay dead. Five
    thousand sixty and three

Was his counting of all their slain. Whilst the
    tale which was brought to me

By En the son of Eschmun was sixteen hundred
    and five,

Nigh the half of whom were miledh. These seek
    not for God to strive,

But for gold and crowns and pillage. Having
    nor child nor wife,

Such lust as steeds after battle, and take a life
    for a life.

Therefore I bade the priests uplift in men's ears
    a song

Of the things which under the Lord should unto
    the queen belong.

(5)   Peace with the Lord *
         The Lord with man

* Literally,   Peace to heaven.
             Heaven to earth.
             Earth under heaven.
             A strength for all peoples.
               See lines on title page and at end.

Man 'neath his Lord
Hath strength to plan.
I would not behold in a wide realm, dear to me
Shame of sisters,
Brothers unbridled,
Seedless summer,
Or plains unpastured.
Captives kingless,
Wise men witless,
Preachers prayerless,
Or any uncleanness.
Rulers unrighteous,
Unjust judges,
Rich men robbers,
Or strong men spoiling.
Undutiful daughters,
Strengthless soldiers.
Betrayers of truth,
And workers of wickedness,
Such will I shame.

Chapter XXIX

(1) *Eocaid after the fight at Magh Tuiread (Moytura, the plain of towers, from the numerous burial heaps there) is wounded well-nigh unto death by Cethlenn the wife of Balor;* (2) *He is healed, yet not to his former might, by Diancecht;* (3) *Tephi, journeying eastward, telleth the shameful death of Crimthann.*

(1) In the midst of mourning, my pride had fall,
    being led astray.
The Lord had lifted me up. The Lord should
    cast me away,
Till my pride was humbled before him. My
    husband, my lover, my friend,
How great that morn was thy strength ; how near
    that eve was thine end.
I sat in my judgment place, and my soul was
    lifted to see
The widow of Balor draw nigh to ask a grace at
    my knee,
Cethlenn,* of evil mouth. Men builded her
    husband's heap,

    * Literally " of the crooked teeth."

And she prayed her burial with him.   Then said
   I, "Ye hold too cheap
My word from the Lord against Baal.   Behold,
   his burnings shall cease.
I will break the horns of his altars, that so my
   people have peace."
Then leapt she upon my side, upraising a little
   knife,
And thrusting it down upon me, thought to have
   had my life ;
But Jochad, springing upon her, lifted her hand,
   and tore
The blade from her grasp, but in struggle, it fell
   and it scratched him sore
By the foot.   Then I bade men take her and
   carry her over sea ;
And thereafter had will to slay her, yet Jochad
   let this not be.
He said how his hurt was little, thus had I comfort awhile ;
But turning my face on my lord for counsel, I
   saw the smile
Die out of his face, and he staggered, for poison
   was in that wound,

And his eyes were darkened before me, and he
stretched himself on the ground.

(2) Six months my watchings endured, and my
sorrow and toil were great,
Ere Diancecth, the mighty healer, cured him, yet
not to the state
Wherein he had strength before.   Of his limb he
was ever lame.
Yet his hurt was healed of the Lord to bring him
a righteous fame,
For he read in the wisdom of God, and drew the
learned in schools,
And taught the scribes till they marvelled.
Moreover he set the rules
Of the three-year meetings at Crofinn, where that
chamber ample and round
Is builded, wherein I will stretch me until my
bones shall be found,
Whensoe'er my White Champion seek me.
There will I dwell alone,
Whilst this land that I builded up by its idols
is overthrown,

And the workings of evil amongst ye.   The
    heathen shall swarm with the waves,
To seek the tombs of my children, and wash
    them out of their graves.
Ernmais and Figol and Elier have counselled of
    this with me.
My tomb shall rest with my people.   Their
    wailing place shall it be
For all that repent them of sin.   Of Ernmais
    the Lord was the eyes,
Yet Jochad had many visions, and therefore men
    called him wise
"Ollam Fothla" the sage of our island, a title
    whereby he is known
Unto many tribes and peoples the furthest from
    Eriu's throne.

(3) In the Springtide, glad at his healing, we journeyed
    out of the West,
With Jochad borne on a litter, and he made his
    chiefest request
That the miledh be given to Lugaid, who went
    not back unto Spain,

But set his hand upon mine, and sware with me
    to remain,

My brother, my champion, my servant. Right
    well hath he kept his word,

Cleansing the woods of robbers, and striking
    down with the sword

All pirates that harried our shores; with the
    vessels of Bres as his aid,

Our hamlets and homesteads had rest, and our
    women walked unafraid.

But now, he would go against Crimthann, and
    therein I answered him "nay,

His faith was broken with David. The Lord is a
    lion in his way."

This was beheld of many, for Crimthann had
    kept the shore,

And guarded our eastward rear to keep by the
    oath he swore;

Yet brake it in working evil, riding for spoil at
    his will.

His mighty men even now were set beside Usna's
    hill;

And there, as he hunted the woods, my complaint
    was heard of the Lord;

For Crimthann, the mighty champion, fell not
    down by the sword

But stoned unto death by swineherds. He had
    cast forth his hunting spear,

And rode alone in the birchgroves to follow a
    wounded deer,

Which fell near the plundered swinepens. Then
    when in his wrath he came

Where the famished swineherds stripped it, they
    rose, and he died in shame.

Then set I his men with the miledh, and Lugaid
    had toil with these,

But, as master of all endeavours, he drew these
    wolves round his knees,

Till they fawned as they fawned not on Crim-
    thann, licking the palms of his hand

For the feastings at Lugaid's table, and his praise
    which was great in the land.

Chapter XXX

(1) *At Tailtea* a firstborn son is given unto Tephi, and she beholdeth the blossom of her seed which she had planted;* (2) *she maketh a confession of sin and its punishment, and admonishes her children thereby, revealing many things unto them.*

(1) At my fortress three months I rested, and a
    strong man-child I bear
To my husband, my firstborn, Aed; now my
    infant was very fair,
Till I loved him more than my land, and my
    heart was severed from God.
The Lord that gave him hath taken. I am sore
    chastised with His rod.
Yet the morn that I carried my firstborn forth
    'neath the summer sky,
How sweet were all scents and sounds, and how
    lovely my land did lie,
For the field was rosy before me that once was
    mantled with green;
And Maistiu, clapping her hands, said, "Praise
    be to thee great queen,

* The strength or stronghold of Teia.

For thou spreadest fair carpets in Eriu, thy
    carpets out of the East

Whereon her children walk softly, her cattle make
    gladdest feast."

In wonder I said, "What mean ye?" She
    answered, "That seed of thine

Thou plantedst last year with care, behold it
    before thee shine

Where it spreadeth on all the field. Thereon do
    thy oxen feed.

It shall grow beside all rivers, for we call it our
    Rigan's seed.*

Now other seeds that I brought from the ships
    had been saved alive.

In my garden of Tailtea I set them, and some
    had the strength to thrive,

Whilst many withered and died. Yet that linen-
    seed, with a flower

Like the heavens, was much increased, till men
    said that the richest dower

Which Tephi brought to the land was seed that
    I plucked by the way

* Clover. See design on cover.

When I went through the grasses from Egypt.
  The Lord was my Stone and my stay
When little I guessed His purpose. Few things
  are yet to be told.
My body is worn and wasted, though by days
  and by years not old,
With long service in aid of this people, in strivings
  and sorrows oft.
Though my love stood by me to ease me, behold
  my couch was not soft.
Our judgments and laws and teachings, are they
  not writ in the book
Of En the scribe and his son, wherein he that
  hath skill may look.
My psalms are laid with the priests. My songs
  do the harpers sing.
May my heartsongs bring cheer to many, my
  psalms find grace with the King,
When I have rest after toiling. Yet one deed
  the Lord hath known,
And two most dear, but in part. This sin of my
  soul will I own
Ere I rest in the hope of Jacob.

Q

(2)　　　　　　　Ye know how I loved my son,
　　My firstborn, believing that he should be mine
　　　　anointed one,
　　Returning in glory to Zion, nay, spake my hope
　　　　unto all.
　　As he dwelt right fair on my bosom.　Ah, why
　　　　must my soul recall
　　His tomb.　I will seek him to aid him.—When
　　　　Ainge my daughter came,
　　I gave her a foster-mother, which thing was often
　　　　my shame.
　　Though she loved me, soon she left me, for a
　　　　husband that deals not well
　　With my Prince, and hath spoiled the trapdams
　　　　he set in the stream to swell
　　Its course ere it passeth seawards ; and cares not
　　　　fresh farms to win
　　From the wolf and the bear, and the bringing of
　　　　sheep and of oxen in.
　　Were he not grandson of Nuadh mine anger had
　　　　been more sore.—
　　Why do I shrink and wander ?　God bids me
　　　　eat to the core

The apple of Sodom I planted.—My third babe
    lay at my side,

Strong and sturdy and fair, yet little in him was
    my pride.

I remembered not how I mourned after love in
    the house of my sire.

My firstborn alone I cherished, till a message
    went forth as fire

From the Lord.   My first born strove in evil rage
    with the Queen,

Who chastised not his froward angers ; whilst
    Angus I had not seen,

But left him in Maistiu's sunhouse,* who ever
    sung by his bed.

Then went I thither and found my blind sister
    with bended head,

Threading a sign on the breast of the babe, and
    I asked her thereof,

For that mark I knew not.   She said, " Many
    righteous his sign shall love,

For deep in the still night watches I heard, as it
    were a voice

---

* Grianan.   The separate house of a woman of rank.

Of one old, compelling mine heart, which said,
   ' Oh virgin, thy choice
With God is seen of His eyes. He giveth into
   thy hand
His token of blessing and sorrow, that thy soul
   may understand
In the dark, and believe His glory. Moreover, it
   shall be set
As a sign on the child thou lovest. Though his
   sorrow cometh not yet,
Nor his blessing till times appointed. Take this
   in thine hands to hold,
Setting lips thereon that it bless thee. Let thy
   fingers veil it with gold,
For a sign unto nations and times that the Branch
   shall ever abide,
Which out of a double thorn is parted on either
   side,
As the props of the Vine I planted.' " Then
   knew I of whom she spake,
And thought of my firstborn, and chideth sore in
   my wrath for his sake,
Then, seizing the four-thorned charm which
   Maistiu had bound with gold,

I broke from my babe its strings, and deep in my
  garment's fold

Bore it swift to his brother ; but the lad in an
  evil mood

Flung it on earth before him, setting his feet on
  the wood,

Which pierced his heel, and he angered, and set
  his teeth to my wrist,

For the serpents arose up in him. * Then lo,
  ere ever I wist

That any man came, one spake, and said, " Wilt
  thou strive with God ?

Thou art even a foolish daughter.   Thou settest
  thy back to the rod.

Thou hast robbed one child of his blessing. Thou
  hast brought his fellow a curse.

Thou knowest the serpents with him.   Thou
  makest their venom worse.

That which thou sparedst to slay, shall sting even
  him and thee

In that day when he doeth great evil.   Then
  truly thy mourning shall be,

---

* Aedh is reported to have had three serpents in him, which
would have destroyed the kingdom of his mother but for his death.

That long time hast not wept for Zion.   Thou
    art proud in thine own estate.
Thine eyes shall be pools of salt, thine affliction
    be very great.
This fourfold thorn shall tear thee.   To thy sister
    make plain thy sin.
David shall come not to Zion till pardon by this
    he win,
And he findeth one pure of heart, and perfect
    before the Lord,
And patient beneath these thorns his city is not
    restored."
Now I lay down under his feet, but saw him
    turning to go,
Whether spirit or man I know not, but he bore
    the mark on his brow
Of that sign, and it shone above me as I lay on
    my face and wept
Long time, whilst Aedh had fled.   Then back to
    Maistiu I crept
With sorrow bound to my heart, and wept on her
    breast and prayed ;
And at morn I bade that a wall by the door of
    my house be made,

Whereon ye have seen me weep over Zion through
   every fast.

Nigh twenty years have I wept, but my weepings
   are overpast;

For I go unto Him that made me. Yet, weep
   ye my children still.

Weep not your mother, but weep over Zion by
   my burial hill.

Tea Mur, my wall, ye shall call it; but David's
   Lord must ye know

If your feet would carry you backwards to con-
   quer his final woe.

I give you words of remembrance, see that the
   same ye bind

On your foreheads to save from idols, and treasure them in your mind.

" Captivity, Bonds, Destruction." *   Keep these,
   being mindful of me,

And this fair isle shall be safe from every robber
   by sea.

Yet these ye will not remember.   I see the ships
   in the bay,

---

* These three words seem to have been so often in Tephi's
mouth, that later bards call Aedh, Angus and Cermad her sons
by them.

When brother slayeth his brother.   Again, I be-
    hold the day

When the Son of Sorrow brings sorrow.   Then
    cometh the bull to gore.

Then my Rock is set upon him.   Behold, I may
    speak no more.

My secret sin is upon me, yet sought I its burden
    might be

Lifted away from my son, and the whole be laid
    upon me.

Ah me, is it three years only ?   It is longer than
    all my life

Since Corrgenn came from his hold to bide near
    us, bringing his wife,

A brother's daughter to Grisbane, and like as the
    twain were twins.

Then our hearth had little honour, and two were
    slain in their sins.

An eric was proffered before us, as for the son of
    a queen,

But Jochad judged that this island were an eric
    all too mean

For me, and for David's heir, if slain in an idle strife.

Yet the Lord of David slew him.   Let Corrgenn
    deal with his wife,

And that other corpse alone.   Betwixt him and
    the Lord these lay ;
And my soul bowed down unto Jochad and rose
    not to say him "Nay."
Therefore Corrgenn bear both unto Ailech, and
    no man went by his side,
And of shame and his toil he turned his face to
    the wall and died,
Leaving his lands and people, and the care of
    that place to me,
So went I forth with my servants Gabhran and
    Imcheal to, see
The grave, and raised up a tomb as they build in
    the land of the Greek,
A rounded chamber of stone that climbeth up to
    a peak
In circles of flags as it narrows, the most fair in
    this land, and alone
Upon Ailech my sins are heavy, and heaped to a
    pillar of stone.
There mine eyes were pools of salt, and also
    Jochad and ye
And the men and babes of my people were one
    in their grief with me.

CHAPTER XXXI

A lamentation of Tephi wherein she giveth instruction.

To be sung to the harp upon the two thousand four hundred and eighty-fourth day.

O, MY CHILD, O, Aedh my firstborn, and O, Aedh my
    firstborn child,

That lay small and warm on my heart and looked in
    mine eyes and smiled

As a flame * thou hast seared my breast, and wert by
    a flame beguiled.

O, fair was my strong son Aedh, and O Aedh, my
    strength, was fair.

The skies were seen in his eyes. The sun was set
    in his hair.

The Mighty hath slain my son. I mourn, yet He
    might not spare.

O, mine eyes are rivers of tears, and O, rivers of tears
    are mine eyes.

I sat in the seat of folly. I walked not amongst the
    wise.

I sowed a seed of destruction. Its fruits are foulness
    and lies

* Aedh, a flame.

O, let evil be upon Canaan, and O, upon Canaan be
 every ill.

Why hale ye their women hither, that are harlots on
 every hill,

That are brazen in dances to Baal, that are wanton in
 all their will ?

O, hear me, my chosen, my husband, and O, my
 husband, my chosen, hear.

I have erred and have done great evil. My burden
 is heavy to bear.

This mocking was mine not thine. Yet my shame
 hath been thine to share.

O, heed me Angus, my son, and O, Angus, my son,
 take heed.

Thy brother is black in the pit. He stinks as a
 rotten reed.

Thou bearest the Branch of blessing. Thy Stone is
 chosen for seed.

Yet I know thee, O, Angus, my son, and O, Angus,
 my son, I know

Thy pomp and thy pride of heart. Thy flame
 burneth on and fro.

It flasheth fire in the sky. Its light is sunken and low.

I divine thee, O Angus, my son, and, O Angus, my
       son, I divine
Thy spirit unscarred by the thorns.   Thou shalt seek
       but the gold of that sign.
Thy heart is not with the High One.   With sinners
       thou sittest at wine.

I behold thy grave,* O, my son, and thy grave, O,
       my son, I behold.
Thy grave-mound is glorious and great.   Thou
       graspest there on thy gold,
Yet the heathen shall find thy hoard ere the hill of
       thy height wax old.

O, thy treasure is heaped upon earth, and O, with
       earth is thy treasure-heap.
Thou art e'en as the kings of Egypt.   Thou sinkest
       down in thy sleep.
But thieves shall find thee therein, and the snail and
       the slow-worm creep.

Thy toiling is waste, O Angus, and, O Angus, waste
       is thy toil.

* Œngus, of the Brugh, is now best remembered by this
enormous tumulus, which was plundered by the Danes.

Thy masons build thee a mansion. The spoiler
    shall make it a spoil,
For thy zeal is not unto Zion, nor thine heart
    anointed with oil.

O, may the bright reign come by thee, and O may
    my white king come.
His sheep he leadeth in spirit. He rebuketh them
    lest they roam.
He blesseth their lambs in his bosom. They hear
    him at eve and go home.

O, hear ye the promise of Israel, and O, Israel, this
    promise hear.
Let your watchmen know of the night. Let them
    count when the stars grow clear.
Let them strongly shout in the gate if a presage of
    dawn appear.

O, rest ye your faith upon David, and O on David
    let fealty rest.
In righteous judgments he rideth. His wise men
    gaze from the west.
His house on the hill-tops is holy. His symbols
    shine on his breast.

O, he rides as a king in glory, and O, in glory my
　　king doth ride.
The nations are scattered beneath him.　In their
　　eyries the eagles hide.
As a lion he leaps in his strength.　What man shall
　　his might abide.

O, springs gush out by the Hill, and O, from the Hill
　　there gush forth springs.
O'er the path of his chosen people, the vessels bear
　　wealth unto kings.
The ships of the sea pass over.　The waters are
　　white with their wings.

O, broad is the stream of Jordan, and O, Jordan thy
　　streams are broad.
The seas have set thee in might.　No steed shall
　　swim by thy ford,
Where the House of the High One is builded, the
　　Holy House of the Lord.

O, now I depart in peace, and O, peace is my part as
　　I go.
I have lived the days of my life.　I have joyed and
　　wandered in woe.

I am feeble and fain would rest from my travelling
to and fro.

But, O, that day I am fain to behold, and O, I fain
would behold that day.
Raise up the stones from my sidhe. Cleanse ye my
bones from the clay.
Let me see the son of my strength, for my spirit
shall be his stay.

CHAPTER XXXII

Garbh Cliach, the recorder, the son of En, writes of that which
may not be written save upon the hearts of the men of
Eriu.

Now the rest of the acts of Teffia, and how her
sunhouse was made
At Tailtea, the beams of its rafters with wings
of bright birds o'erlaid,
And its hurdles snow under summer, so that
men's eyes were blind
Beholding, and how its porches with plates of
silver were lined;

And her purple couches within; and her crowns
and bracelets of gold,

That often she gave to the bards; and the things
which her shipmen sold

In her mart; and the peace and joy of her land;
and her two fair sons,

Œngus the frank and Cermad; and the many
cashels and duns

She set for defence of the sea-coast; and the
mighty forests she cleared;

And her wide ensample to all men; and the
grace that in her appeared

Before kings and sages and lowly (for of all men
her speech was known

As a dew that falleth from heaven, and holy
before God's throne,

Yet was troubled in many sorrows alike of bonds-
man and free;)

And how in Crofinn a house was built that her
rest might be

Beside the assemblies of Eriu to soften their
judgments still,

And stay their sharpness of strife 'neath the shade
of the Great Queen's hill;

And how she had many champions and bards
  and sages and priests ;

And how men wise in the Lord came from afar
  to her feasts ;

And how many kings sent greetings ; and how
  she was mourned for and wept

Through the whole green isle of Eriu, and women
  came where she slept,

Yea, e'en from the utmost islands to shed on her
  sidhe their tears,

And planted their flowers about it ;—It needs
  not that aught appears

In the books of the scribe, for all is written large
  on the heart

Of Eriu, although she oft told presage her name
  should depart

From our lips for a season, if these by her psalms
  be not purified ;

And that if men failed of her trust, her blessing
  should be denied ;

Yet, know we well that her blessing shall ne'er
  be taken away,

Nor her face be ever hidden, although it be veiled
  for a day.

R

So also the Heremon liveth, though under his
    stones he lie
On the hills * o'er the lake, his glory and honour
    shall never die
Of bard and champion and teacher and lifter of
    burdens sore,
Which against the might of his word the hands
    of his sons restore ;
Till the Firbolgs toil, as in Egypt our fathers
    were wont to toil,
On the tombs that they build by Boyne, filling
    their pouches with soil
To heap on the secret chambers wherein these
    would build their home
At the last ; and thither surely their bones with
    the curse shall come
Of our loved one † and not her blessing. Also
    men have much grief
Against Ethdan grandson of Nuadh, whom the
    unwise chose as their chief
Of the miledh after Lugaid, for he taxeth the
    land of its yields

* The Loughcrew Hills.

† Tephi is alluded to merely as "the Beloved" in early
documents.

Beyond the strength of the aire, and letteth the
    woods on their fields;

And save that Ainge, his wife, is loved of the
    people still,

As the child of our Ollam Fothla, some surely
    had wrought him ill.

Though the bards sing many complaints, the
    princes repent no whit,

Therefore Garbh, the son of En the son of
    Eschmun, hath writ

These words in this book against them. For our
    evils will never cease,

Till the word of Tephi prevail, and her last and
    her foremost was " *Peace.*"

Peace unto God in heaven. Let God shine
    thence upon earth,

And the Branch shall anoint you with oils of
    blessing and praise and mirth.

            Sith co Nem
            Nen co Doman
            Doman fo Nim
            Nert hi cach.

FINIS.

www.ingramcontent.com/pod-product-compliance
Lightning Source LLC
Chambersburg PA
CBHW021517210326
41599CB00012B/1284

* 9 7 8 3 7 4 3 4 7 1 8 7 0 *